Mastering jQuery Mobile

Design and develop cutting-edge mobile web
applications using jQuery Mobile to work across
a number of platforms

Chip Lambert

Shreerang Patwardhan

BIRMINGHAM - MUMBAI

Mastering jQuery Mobile

First published: July 2015

Production reference: 1270715

Published by Packt Publishing Ltd.
Livery Place
35 Livery Street
Birmingham B3 2PB, UK.

ISBN 978-1-78355-908-4

www.packtpub.com

Credits

Authors
Chip Lambert
Shreerang Patwardhan

Reviewers
Saeed Afzal
Altaf Hussain
Anirudh Prabhu

Commissioning Editor
Owen Roberts

Acquisition Editor
Tushar Gupta

Content Development Editor
Adrian Raposo

Technical Editors
Novina Kewalramani
Shiny Poojary
Mohita Vyas

Copy Editors
Puja Lalwani
Kausambhi Majumdar
Sameen Siddiqui
Ameesha Smith-Green

Project Coordinator
Kinjal Bari

Proofreader
Safis Editing

Indexer
Tejal Soni

Graphics
Sheetal Aute
Disha Haria
Jason Monteiro

Production Coordinator
Manu Joseph

Cover Work
Manu Joseph

About the Authors

Chip Lambert has been creating websites ever since his high school days, when he started with fantasy sport websites. In addition to these 20 years of HTML experience, he has 15 years of experience with PHP and MySQL development. He previously wrote *Instant RESS Implementation How-to* by Packt Publishing. He is currently a technical analyst for Jenzabar, Inc. You can follow him on Twitter at `@chiplambert` or visit his personal blog cum website at `http://www.programmerchip.com`.

First, I would like to thank my wife and daughter for their patience and love during the writing of this book. I love you both very dearly. I would like to thank my parents for everything they've done over the growing-up years to keep me alive.

Also, a big thanks to Shreerang for coming in at a difficult time and helping out with the completion of this title. Also, thanks to all the technical reviewers for their help.

Shreerang Patwardhan completed his bachelor's degree in computer engineering, and since then, he has been working on various technologies for the last 4.5 years. He started off by working for a small start-up in Pune, India, on an innovative solar-powered Linux-based handheld device. He has also worked extensively on Google Maps API v3 and worked in the GIS domain for more than a year. He is also one of the technical reviewers for *Google Maps JavaScript API Cookbook*, published by Packt Publishing.

Shreerang is currently employed at a MNC in the San Francisco Bay Area, USA, as a technical consultant and is working on the frontend development of various web applications using different cutting-edge frontend technologies. He is also a certified Scrum Master and absolutely loves and encourages the Agile methodology.

When not working on a project, he blogs about either the Google Maps API v3 or the jQuery Mobile framework on his blog, *Spatial Unlimited*. When not working or blogging, Shreerang loves spending time with his family and friends and absolutely enjoys sweating it out on the badminton court. He has been playing badminton for the last 20 years and also takes a keen interest in UFOlogy.

You can reach him on his blog, contact him on LinkedIn, or follow him on Twitter (@shreerangp).

I would like to take this opportunity to thank the people who have directly or indirectly played a huge role in me authoring this book. First and foremost, I would like to thank Amit Karpe, my first reporting manager and a dear friend, who introduced me to writing and always encouraged me to write a book and had faith in my capabilities.

Thanks to Adrian Raposo, the content development editor for this book, for believing in me, and for providing me with the opportunity to author this book with Chip. Thank you for all the patience and encouragement that you provided us during the process of writing the book.

A huge thanks to my colleagues from various organizations, from whom I have learned a lot technically and continue to learn, especially the developer friends, seniors, and peers from Cybage Software, Pune, India. Thanks to all the folks from this organization for always being available to help and guide.

Thanks to my friends, who have always been there for me and tolerated my madness.

Finally, thanks to my parents and my brother, who have always backed me in whatever I pursued and have always believed in me and encouraged me to pursue my dreams.

About the Reviewers

Saeed Afzal is known as Smac Afzal. He is a professional software engineer and technology enthusiast based in Pakistan. He specializes in solution architecture and the implementation of scalable high-performance applications.

He is passionate about providing automation solutions for different business needs on the Web. His current research and work includes the futuristic implementation of a next-generation web development framework that reduces the development time and cost, and delivers productive websites with many necessary and killer features by default. He is expecting the launch of his upcoming technology in 2016.

He has also contributed to the book *Cloud Development and Deployment with CloudBees, Packt Publishing*.

More detailed information about his skills and experience can be found at `http://sirsmac.com`. He can be contacted at `sirsmac@gmail.com`.

I would like to thank the Allah Almighty, my parents, and my wife, Dr. H Zara Saeed, for encouragement.

Altaf Hussain is an electrical engineer on paper and a software engineer at heart. He is an e-commerce and mobile applications enthusiast. He has a bachelor of engineering degree in electrical engineering (specialized in computer and communication) from Pakistan. He worked in numerous organizations as a backend developer and then moved to Saudi Arabia as a software engineer.

Currently, Altaf is working in the fashion industry at shy7lo.com. He is managing dedicated servers, different VPSs, staging servers, and gitlab instances for fast deployment. As a senior team member, he is responsible for creating cross-platform mobile applications and APIs. Also, he is working on different caching systems, such as Varnish and Full Page Cache. In his free time, Altaf writes posts for http://www.programmingtunes.com/.

> I would like to thank and congratulate the author for writing this amazing book. This is a must-have book for frontend and mobile app developers.

Anirudh Prabhu is a software engineer (Web) with more than 5 years of industry experience, specialized in technologies such as HTML5, CSS3, PHP, jQuery, Twittter Bootstrap, and SASS. Also, he has entry-level knowledge of CoffeeScript and AngularJS.

In addition to web development, he has been involved in building training material and writing tutorials for http://www.twenty19.com/ on the following topics:

- Learning to build websites with HTML and CSS
- An introduction to jQuery

He has been associated with Apress and Packt Publishing as a technical reviewer for several of their titles.

www.PacktPub.com

Support files, eBooks, discount offers, and more

For support files and downloads related to your book, please visit www.PacktPub.com.

Did you know that Packt offers eBook versions of every book published, with PDF and ePub files available? You can upgrade to the eBook version at www.PacktPub.com and as a print book customer, you are entitled to a discount on the eBook copy. Get in touch with us at service@packtpub.com for more details.

At www.PacktPub.com, you can also read a collection of free technical articles, sign up for a range of free newsletters and receive exclusive discounts and offers on Packt books and eBooks.

https://www2.packtpub.com/books/subscription/packtlib

Do you need instant solutions to your IT questions? PacktLib is Packt's online digital book library. Here, you can search, access, and read Packt's entire library of books.

Why subscribe?

- Fully searchable across every book published by Packt
- Copy and paste, print, and bookmark content
- On demand and accessible via a web browser

Free access for Packt account holders

If you have an account with Packt at www.PacktPub.com, you can use this to access PacktLib today and view 9 entirely free books. Simply use your login credentials for immediate access.

Table of Contents

Preface

jQuery Mobile really ups the ante for us mobile developers. It gives us a powerful framework to develop web applications tailored for mobile devices and have them act like native applications. We hope to take your jQuery Mobile skills to the next level with this book by having you develop projects that you may encounter in your everyday life, instead of a bunch of one-shot conceptual projects. We have one main project that we will build over a series of chapters and then some standalone projects that will explore some more advanced features.

What this book covers

Chapter 1, Getting Started, in this chapter you will begin setting up your development environment to be able to complete the project accompanying the book. You will learn the various methods to obtain and use the framework in a small sample application.

Chapter 2, Tools and Testing, in this chapter you will look at some additional tools that you can use in jQuery Mobile development. Apart from the different tools, this chapter will also look at some means of testing that will assist you in making sure your projects work well across all platforms.

Chapter 3, Mobile Design, in this chapter you will get a brief overview of developing for mobile devices. The topics will cover responsive web design techniques, including media queries, features, and device detection, and will take a very brief look at RESS.

Chapter 4, Call to Action – Our Main Project, in this chapter you will begin the development of our overall project, a mobile application for a fictional Anytown Civic Center. The application will display a list of upcoming events, allow users to enquire for the events, and serve as a mobile website for the civic center itself.

Chapter 5, Navigation, with this chapter, you will start putting together a touch-based navigation system as we link the pages together. You will also get to see the use of transitions.

Chapter 6, Controls and Widgets, in this chapter, you will explore the various widgets and controls that we will use to build the Anytown Civic Center application. You will learn how to configure and initialize widgets, and control the app with various input events.

Chapter 7, Working with Data, in this chapter, you will look at how we can retrieve data using PHP and use it within our mobile application. We will look at forms, retrieving data from a database, validating input before inserting it back into the database, and displaying information to the application user.

Chapter 8, Finishing Touches, in this chapter, you will finish up the Anytown Civic Center mobile application by creating custom icons to use in the application to replace the default ones. We will also begin exploring how this application can be turned into a native mobile application with a basic introduction to Cordova.

Chapter 9, The Next Level, in this chapter, you will look at using jQuery Mobile with the Node.js platform, integrating jQuery Mobile with the Backbone.js and RequireJS libraries, and build a basic WordPress mobile theme that will use this framework.

Chapter 10, Mobile Best Practices and Efficiency, in this chapter, you will look at some best mobile practices and some optimization tips and tricks that will benefit you as you go deeper into jQuery Mobile development.

What you need for this book

For you to get the most out of this book, we recommend you to go along with the projects and code, so for this, you would need to have access to some sort of development device, be it a laptop or desktop. We recommend these, so that you can get the most out of the book by developing locally. You could, however, do this from your tablet device as well using one of the cloud based IDE systems such as jsFiddle, Codeanywhere, and so on. We will walk you through setting up a powerful and free environment with Aptana Studio and XAMPP.

Who this book is for

You've started down the path of jQuery Mobile, done a small application using some basics of the framework. Now, if you wish to master some of jQuery Mobile's higher level advanced topics, you need not look further. Go beyond jQuery Mobile's documentation and master one of the hottest mobile technologies out there. Previous JavaScript and PHP experience can help you get the most out of this book.

Conventions

In this book, you will find a number of text styles that distinguish between different kinds of information. Here are some examples of these styles and an explanation of their meaning.

Code words in text, database table names, folder names, filenames, file extensions, pathnames, dummy URLs, user input, and Twitter handles are shown as follows: "Double-click on the .DMG file you just downloaded and drag the XAMPP folder into the Applications directory."

A block of code is set as follows:

```
<link rel="stylesheet" href="http://ajax.googleapis.com/ajax/libs/
jquerymobile/1.4.5/jquery.mobile.min.css" />
<script src="http://ajax.googleapis.com/ajax/libs/jquery/1.11.0/
jquery.min.js "></script>
<script src="http://ajax.googleapis.com/ajax/libs/jquerymobile/1.4.5/
jquery.mobile.min.js"></script>
```

When we wish to draw your attention to a particular part of a code block, the relevant lines or items are set in bold:

```
<p>Welcome to the website for Anytown Civic Center. We have a lot of
great upcoming events for you to check out.</p>
```

New terms and **important words** are shown in bold. Words that you see on the screen, for example, in menus or dialog boxes, appear in the text like this: "Double-click on the installation file and you will be asked to select a language. Make your choice and choose **OK**."

 Warnings or important notes appear in a box like this.

Tips and tricks appear like this.

Reader feedback

Feedback from our readers is always welcome. Let us know what you think about this book—what you liked or disliked. Reader feedback is important for us as it helps us develop titles that you will really get the most out of.

To send us general feedback, simply e-mail feedback@packtpub.com, and mention the book's title in the subject of your message.

If there is a topic that you have expertise in and you are interested in either writing or contributing to a book, see our author guide at www.packtpub.com/authors.

Customer support

Now that you are the proud owner of a Packt book, we have a number of things to help you to get the most from your purchase.

Downloading the example code

You can download the example code files from your account at http://www.packtpub.com for all the Packt Publishing books you have purchased. If you purchased this book elsewhere, you can visit http://www.packtpub.com/support and register to have the files e-mailed directly to you.

Errata

Although we have taken every care to ensure the accuracy of our content, mistakes do happen. If you find a mistake in one of our books—maybe a mistake in the text or the code—we would be grateful if you could report this to us. By doing so, you can save other readers from frustration and help us improve subsequent versions of this book. If you find any errata, please report them by visiting http://www.packtpub.com/submit-errata, selecting your book, clicking on the **Errata Submission Form** link, and entering the details of your errata. Once your errata are verified, your submission will be accepted and the errata will be uploaded to our website or added to any list of existing errata under the Errata section of that title.

To view the previously submitted errata, go to https://www.packtpub.com/books/content/support and enter the name of the book in the search field. The required information will appear under the **Errata** section.

Piracy

Piracy of copyrighted material on the Internet is an ongoing problem across all media. At Packt, we take the protection of our copyright and licenses very seriously. If you come across any illegal copies of our works in any form on the Internet, please provide us with the location address or website name immediately so that we can pursue a remedy.

Please contact us at copyright@packtpub.com with a link to the suspected pirated material.

We appreciate your help in protecting our authors and our ability to bring you valuable content.

Questions

If you have a problem with any aspect of this book, you can contact us at questions@packtpub.com, and we will do our best to address the problem.

1
Getting Started

Are you all excited, as we are, to get started? Well, before we start writing some code, we have a few things we need to do first. These first couple of chapters may be light on code, but are rich in concepts and ideas that you will need on your journey to master jQuery Mobile.

Before we begin, let's talk briefly about what **jQuery Mobile** is and, more importantly, what it isn't. It is an extension of jQuery that will allow you to create cross-browser web applications and websites that will look the same, regardless of the mobile browser used to view them and the underlying mobile OS. jQuery Mobile provides wide browser and device coverage using progressive enhancement, making your website accessible on the widest range of devices and browsers. It also enhances the form inputs and UI widgets to be touch-friendly. It is not a framework that will allow you to build native applications, but you can use this framework to build hybrid mobile applications. In one of the last chapters of this book, we will look at using one such framework—Apache's **Cordova**.

Now that we have got that out of the way, grab your drink of choice, fire up your computer, and let's rock 'n' roll.

Overview

In this chapter, we will look at installing the XAMPP server stack for both Windows and Mac OS X, installing the Aptana Studio **integrated development environment (IDE)**, and then start looking at using jQuery Mobile and the various methods you can use in your project to include the framework. We will then create a small application to get our feet wet with jQuery Mobile. For some of you this could be a simple refresher, but for others this could be a crash course introduction.

You may be asking yourself, "Why am I installing a full server stack, and is jQuery Mobile usage tied to PHP or any of the technology in the XAMPP stack?" While this book is focused on development with jQuery Mobile, we will be doing a very small bit of PHP with MySQL, just to show the interaction between the two, and so we will be setting up our XAMPP stack. However, the usage of jQuery Mobile is completely independent of PHP and any other technology in the XAMPP stack, so developers should feel free to use any other server stack that they prefer for development. If you are not a PHP pro, don't worry; the code will be basic and we will explain it all. If you wish to enhance your PHP skills, check out any of the number of PHP titles Packt has published.

Also by no means are you bound to using your local machine. If you have a web server, feel free to use it; we just find it easier to develop on a local machine so we don't have to worry about making a change, uploading, testing, and repeating.

Installing XAMPP for Windows

XAMPP is a fully functional server stack by the folks at Apache Friends, which includes Apache, PHP, MySQL, Perl, Tomcat, phpMyAdmin, and much more in one installation package. By installing it, you don't have to worry about manually setting up each component of a WAMP stack, you simply execute the installer and choose the components you wish to install.

To get started, we will need to download the latest stable XAMPP installer from `https://www.apachefriends.org/index.html`. We recommend using the installer rather than the ZIP file method for this book.

After we have it downloaded, the installation is pretty straightforward:

1. Double-click on the installation file and you will be asked to select a language. Make your choice and choose **OK**.

2. If you are using Windows Vista or higher, and **User Account Control (UAC)** is enabled, you will see a warning that some functionality may be limited if installed to the `Program Files` directory. Click **OK**.

3. Another warning may pop up prompting you to install the Microsoft Visual Studio 2008 redistributable package. Click **Yes** and you will be taken to a Microsoft downloads page where you can download this redistributable from. Download it and install it before continuing with the XAMPP installation.

4. Click on **Next** and you will be shown all the components that make up the XAMPP package. Select the features you want but for this book, be sure to pick at least **Apache**, **PHP**, **MySQL**, and **phpMyAdmin**, as they will be utilized at some point in this book.

5. The default directory should be fine, so leave it as is and click **Next**.

6. This screen will ask you if you wish to learn more about Bitnami; this is your choice, so if you do not want to see more information, simply uncheck the checkbox and click **Next**.

7. The installation itself will take several minutes to complete. Once it is finished, click on **Finish** from the screen that pops up.

8. You will now see a screen that will inform you that all the services can be controlled through the XAMPP control panel. It will ask you if you want to open that screen now; click on **Yes**.

9. In the control panel, you will be able to start, stop, and restart all the services that were installed with XAMPP. By default, some of these services may be turned off. For now, let's make sure that Apache and MySQL are both running by clicking on **Start** next to the service name.

Installing XAMPP for Mac OS X

XAMPP also has a package available for those running Mac OS X and it is a million times easier to install than its Windows counterpart.

Download XAMPP from the same page as the Windows binaries: https://www.apachefriends.org/index.html. Double-click on the .DMG file that you just downloaded, and drag the XAMPP folder into the Applications directory. After this is completed, open the XAMPP control panel found in Applications and start the Apache and MySQL services.

Regardless of what OS you installed XAMPP, we can verify the installation on both in the same way. Open up your browser and go to the address `http://localhost/xampp`. If all went well, you will see a screen similar to the following:

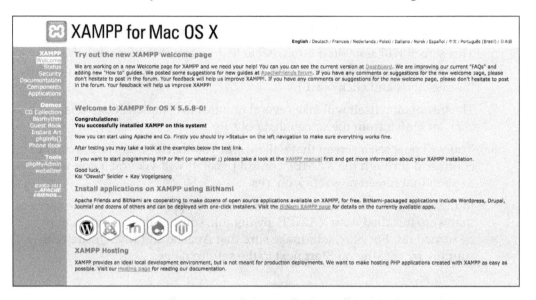

Congratulations! You now have a full-blown testing/development server ready for use to go with this book. If you don't see a screenshot like the preceding one, double check to make sure you did not miss a step.

Important XAMPP Notes

There are a couple things you need to be aware of with regards to XAMPP. First off, it is a great product for situations like this when you need to set up a quick development environment, but it should be used for just that: a development environment. You should not use XAMPP as a production server. Instead, you should install and configure each component individually for various security reasons.

XAMPP for Mac is still listed as being in beta, therefore you could come across some bugs. If running beta software on your Mac scares you a bit, there are other alternatives such as MAMP, which is a very popular Mac stack available at `http://www.mamp.info/en/index.html`.

There is also a version of XAMPP for Linux, so if that is your OS of choice, feel free to download and install it for the purposes of this book.

Installing Aptana Studio

Aptana Studio is a free, open source IDE built off of Eclipse. We will be using Aptana Studio 3 for this book and you can download it from `http://www.aptana.com/products/studio3/download.html`.

To install Aptana Studio 3 on a Windows machine, follow these instructions:

1. Double-click on the installer file and select **Next** to get the installation started.

2. Choose **I Agree** to accept the license agreement.

3. The default directory should be fine, as should the default programs group, so hit **Next** on these two steps.

4. The installer will now prompt you to choose which file extensions Aptana should be the default editor for. Leave the default settings on this screen. You can change them later in the IDE settings if you need to. Click on **Next**.

5. Click on **Install** to start the installation. Once the installation is finished, click on **Next** and then on **Close**.

Installing Aptana Studio on Mac is just as easy as it was for installing XAMPP. After you download the `.DMG` file, open it up and drag the indicated folder into your `Applications` folder.

Now that Aptana Studio is installed, open it up and you will be presented with a dialog box asking you to set up your default workspace directory. Go ahead and set this to be your `htdocs` folder of your XAMPP installation. For you folks on Windows, that directory will be `c:\xampp\htdocs` (provided you used the default directory). For the Mac users, this will be `/Applications/XAMPP/htdocs`. By setting this default workspace, any new projects we create will automatically be created and saved here, making it easier for us to test.

Also for you Mac users, there are some further steps that you may need to do for Apatana Studio to work correctly. By default, the XAMPP `htdocs` folder is owned by root, and because of this, it is set to read-only. In order for you to create and save to this directory, we will need to change those permissions.

1. Click on the folder in **Finder** and then choose **Get Info** from the **File** menu.

2. Click on the triangle next to **Sharing and Permissions** to display the permissions for the selected file or folder.

3. Click on the lock and authenticate with an administrator account.

4. Use the menus next to users and groups to change the permissions.

5. Once you are finished, close the **Info** window.

Of course, if you need further assistance with Aptana Studio or even XAMPP, refer to their respective websites for support. If you happen to fall in love with Aptana Studio and wish to learn more about it, check out *Aptana Studio Beginner's Guide, Thomas Deuling, Packt Publishing*.

> We chose to use Aptana Studio 3 as an IDE for this book, so that you can follow along without having to purchase the same IDE we use (we use Zend Studio). Aptana Studio is free so if you have never used an IDE before, you now have an option. If you have an IDE already that you use on a daily basis, use it for the book. Just adjust any steps listed for Aptana to fit your own IDE.

Downloading jQuery Mobile

Now we finally get to the meat and potatoes of what this book is about: jQuery Mobile. First thing we need to do is download the framework to use it in our projects. You have two ways of doing this. You can download a prebuilt ZIP file containing the latest stable release and everything you will need to use jQuery Mobile. The second option is that you can build a package that contains only the pieces you need, including earlier versions or unstable branches.

In this book, we will be using the complete package. There are couple reasons for this. First of all, we will be using and looking at so many different aspects of the framework that it would take a lot of space here for us to tell you which pieces to select in the builder. More importantly, as noted on the jQuery Mobile website, the builder is currently in alpha and is not recommended for production environments. Yes, we realize that we are not in a production environment here, but we don't want to try to master a framework by using code that could be unstable. We believe that's enough babbling, let's start downloading!

Go to `http://jquerymobile.com`. On the top right- hand side of the page, you will see the links to download the latest stable version (which at the time of writing this book is 1.4.5) and a link to the download builder. Click on the link for the latest stable. After the ZIP file is downloaded, go ahead and open it up and you should see the following:

▶ demos	Today, 2:02 PM	--	Folder
▶ images	Today, 2:02 PM	--	Folder
jquery.mobile-1.4.5.css	Oct 31, 2014, 1:33 PM	240 KB	CSS
jquery.mobile-1.4.5.js	Oct 31, 2014, 1:33 PM	466 KB	JavaScript
jquery.mobile-1.4.5.min.css	Oct 31, 2014, 1:33 PM	207 KB	CSS
jquery.mobile-1.4.5.min.js	Oct 31, 2014, 1:33 PM	200 KB	JavaScript
jquery.mobile-1.4.5.min.map	Oct 31, 2014, 1:33 PM	236 KB	Unix Executable File
jquery.mobile.external-png-1.4.5.css	Oct 31, 2014, 1:33 PM	122 KB	CSS
jquery.mobile.external-png-1.4.5.min.css	Oct 31, 2014, 1:33 PM	91 KB	CSS
jquery.mobile.icons-1.4.5.css	Oct 31, 2014, 1:33 PM	129 KB	CSS
jquery.mobile.icons-1.4.5.min.css	Oct 31, 2014, 1:33 PM	127 KB	CSS
jquery.mobile.inline-png-1.4.5.css	Oct 31, 2014, 1:33 PM	149 KB	CSS
jquery.mobile.inline-png-1.4.5.min.css	Oct 31, 2014, 1:33 PM	118 KB	CSS
jquery.mobile.inline-svg-1.4.5.css	Oct 31, 2014, 1:33 PM	227 KB	CSS
jquery.mobile.inline-svg-1.4.5.min.css	Oct 31, 2014, 1:33 PM	196 KB	CSS
jquery.mobile.structure-1.4.5.css	Oct 31, 2014, 1:33 PM	91 KB	CSS
jquery.mobile.structure-1.4.5.min.css	Oct 31, 2014, 1:33 PM	69 KB	CSS
jquery.mobile.theme-1.4.5.css	Oct 31, 2014, 1:33 PM	20 KB	CSS
jquery.mobile.theme-1.4.5.min.css	Oct 31, 2014, 1:33 PM	12 KB	CSS

That's quite daunting, isn't it! No worries at all though; we will only be referencing a few of the files here, but let's talk about the overall contents of the ZIP file so that you will have a better understanding of what you just downloaded:

- `demos`: We highly recommend that you look at and play around with the contents of this folder. You will find numerous examples here of the different elements of jQuery Mobile. Go now, spend some time with it and then come back—we'll wait.

- `images`: This directory contains all of the icons that jQuery Mobile uses throughout the framework. You won't have to reference these directly, rather jQuery Mobile does so through CSS and its core JavaScript files.

- `*.css files`: These are several style sheets that make up the overall jQuery Mobile framework. They control the preceding icons, the theme (more on that later), structure, and more.

- `jquery.mobile-1.4.5.css`: This is the main and most important CSS file of the download and the one that you will reference and use in your applications.

- `jquery.mobile-1.4.5.js`: This is the core JavaScript of the framework.

You may have noticed that there are two versions of each file. One that has .min appended to the filename and one that does not. Those with .min are compressed minified files. This means they do not have any white spaces, new line characters, and comments in them. They are the absolute smallest in size that they can be and are hard to read as they do not have any formatting. Uncompressed files, or those without .min, contain comments, spacing, and more things that make them easier to read.

As a rule of thumb, you should develop with the uncompressed files; that way, if you have any questions or need to see how something works, you can easily read the file. When you deploy the application, use the .min files so that the application size is smaller.

A smaller web application will load faster, uses less bandwidth, and any little bit helps in a time where we have metered bandwidth on mobile devices. Google checks for the page load time as part of SEO. Faster loading JavaScript and CSS files will help you score brownie points and hence minified (.min) files should be preferred in the production environment.

Using the framework via the CDN

There is a third way to acquire and use the framework. In the previous section, we looked at two different ways in which we can download and include the framework locally in our project. In addition to downloading jQuery Mobile, you can also use it via a **content delivery network (CDN)**.

When using a CDN such as Google or even jQuery's own, you do not have the files stored on your local server. Instead, when the user loads your web application or site, it makes a connection to the CDN and uses it from there. This offers a few advantages over having it on the server. One of the biggest advantages is that it increases parallelism. Typically your browser can only download a couple of files at a time from the same server. So if you load the jQuery files, jQuery Mobile framework JavaScript and CSS files, a few custom CSS, and JavaScript files, this can cause a bottleneck as you will have to wait for all the files to download. If you use a CDN, your application makes a connection to Google's server, for example, and downloads a couple of files while your local ones load. This may not seem like a lot of benefit since the files are so small, but it can make a world of difference to your end user. Some other advantages are as follows:

- If the user visits other sites that use the same CDN, the framework files could already be cached, thereby decreasing your load time even more.

- It reduces the load on your web server. If you have a heavily used site or application, this could decrease the use of your own server resources.

- Using a CDN can also allow the user to download the framework files even faster since there may be a CDN mirror closer to the user location than your web server.

So with all these cool advantages, why in the world would we ever want to use it locally? I mean, it sounds like we'd be stupid to ever download it again. Well, my friends, as with many things in life, there is always a downside.

Obviously, the biggest downfall is offline applications. If you are expecting or even adding the slightest bit of offline functionality to your project, then you cannot have the project going off server to use the framework from a CDN. It would have to be stored locally. If the user is offline, they won't be able to download it from the CDN and therefore would not be able to use the project that you worked so hard on.

Okay, so you're not planning on any offline functionality of your project. What else could stop you from using it via a CDN? Well, my friends we have a few more downfalls:

- If your project is on a company intranet, there may be ACLs or firewalls in place that could prevent your internal webserver from accessing the outside world.

- If your website is using an SSL certificate, the user will get a warning that they are accessing information from an unsecure site. This could spook many users and they may not trust your website.

- If you make custom changes or add new functionality to the framework, you will obviously need to store it locally unless you're an absolute rock star and get your changes rolled into the main builds.

Well, we assume that we have covered the advantages and disadvantages of the usage of CDN well enough, so let's look at how and from where we can load jQuery Mobile.

There are three main CDNs we can use in our projects. They are Google, Microsoft, and of course, jQuery itself. Are there advantages to using one over the other? Possibly, but I think the differences would be small. Google may offer more mirrors but more people may use jQuery's own CDN meaning more people would have it cached, so in the end there may be no difference in efficiency.

We will look at the code for each of the CDNs and you can pull it from whichever one you wish to use.

Google's CDN

Google's CDN code is as follows:

```
<link rel="stylesheet" href="http://ajax.googleapis.com/ajax/libs/
jquerymobile/1.4.5/jquery.mobile.min.css" />
<script src="http://ajax.googleapis.com/ajax/libs/jquery/1.11.0/
jquery.min.js "></script>
<script src="http://ajax.googleapis.com/ajax/libs/jquerymobile/1.4.5/
jquery.mobile.min.js"></script>
```

Microsoft's CDN

Microsoft's CDN code is as follows:

```
<link rel="stylesheet" href="http://ajax.aspnetcdn.com/ajax/jquery.
mobile/1.4.5/jquery.mobile-1.4.5.min.css" />
<script src="http://ajax.aspnetcdn.com/ajax/jQuery/jquery-1.11.0.min.
js"></script>
<script src="http://ajax.aspnetcdn.com/ajax/jquery.mobile/1.4.5/
jquery.mobile-1.4.5.min.js"></script>
```

jQuery's CDN

jQuery's CDN code is as follows:

```
<link rel="stylesheet" href="http://code.jquery.com/mobile/1.4.5/
jquery.mobile-1.4.5.min.css" />
<script src="http://code.jquery.com/jquery-1.11.0.min.js"></script>
<script src="http://code.jquery.com/mobile/1.4.5/jquery.mobile-
1.4.5.min.js"></script>
```

So what are we doing in each of these blocks? Well, the first line of each block `<link rel...>` is loading in the CSS file we need for the project. The first `<script src...>` loads jQuery itself into our project. With the current version of jQuery Mobile, you can use jQuery versions 1.8 to 1.11 or 2.1. You will have to make sure that for every jQuery Mobile application, you will need to refer to the jQuery library before you refer to the jQuery Mobile JavaScript file.

> This brings up a good point: even though we are using jQuery Mobile, we still need to load jQuery core. jQuery Mobile is not a replacement for jQuery proper but more of an extension, much like jQuery UI is.

The final `<script src...>` file loads jQuery Mobile itself. So let's get our feet wet in code now.

Seeing the framework in action

So far in this chapter we have done nothing but talk about jQuery Mobile, how we get it, why we should do things a certain way, blah blah. You're a developer; you want to see this in action and now you will. We are going to make a simple jQuery Mobile application. We'll see the code and then we'll explain what is going on within each piece of code:

1. Go ahead and fire up Aptana Studio 3.
2. Navigate to **File** | **New** | **Web Project**.
3. Pick the default template
4. Name the project whatever you like.

Don't worry too much about everything in Aptana Studio right now, as we'll look at it in more detail in the next chapter.

Right-click on your project and navigate to **New** | **File** from the pop-up menu. Name the file `index.html`. Add the following code to the file:

```
<html>
  <head>
    <meta name="viewport" content="width=device-width,minimum-scale=1.0,maximum-scale=1.0,user-scrolable=no">
      <title>jQuery Mobile: Getting Started</title>
      <link rel="stylesheet" href="http://code.jquery.com/mobile/1.4.5/jquery.mobile-1.4.5.min.css" />
      <script src="http://code.jquery.com/jquery-1.11.0.min.js"></script>
      <script src="http://code.jquery.com/mobile/1.4.5/jquery.mobile-1.4.5.min.js"></script>
  </head>
  <body>
    <div data-role="page">
      <div data-role="header">
        <h3>Web Development Books</h3>
      </div>
      <div role="main" class="ui-content">
        <ul data-role="listview" data-inset="true" data-divider-theme="b">
          <li data-role="list-divider">JavaScript</li>
```

```
        <li>
          <a href="#">Object Oriented JavaScript</a>
        </li>
        <li>
          <a href="#">JavaScript Unit Testing</a>
        </li>
        <li data-role="list-divider">PHP</li>
        <li>
          <a href="#">RESTful PHP Web Services</a>
        </li>
        <li>
          <a href="#">Instant RESS</a>
        </li>
      </ul>
    </div>
  </div>
 </body>
</html>
```

When we execute this code, open up your browser and you should see the output similar to what we will see in the following screenshots on a mobile device:

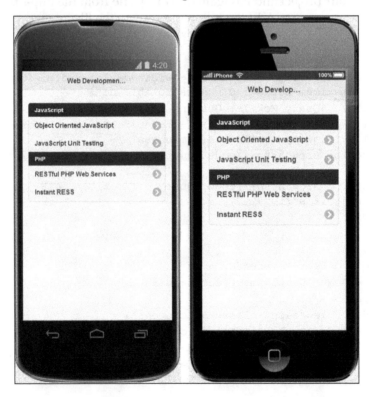

That wasn't so hard, was it? Now, it's time to break down the code and see what's going on under the hood:

```html
<html>
  <head>
    <meta name="viewport" content="width=device-width,minimum-
scale=1.0,maximum-scale=1.0,user-scrolable=no">
      <title>jQuery Mobile: Getting Started</title>
      <link rel="stylesheet" href="http://code.jquery.com/
mobile/1.4.5/jquery.mobile-1.4.5.min.css" />
      <script src="http://code.jquery.com/jquery-1.11.0.min.js"></
script>
      <script src="http://code.jquery.com/mobile/1.4.5/jquery.mobile-
1.4.5.min.js"></script>
  </head>
```

This is your standard HTML5 header with us loading the jQuery Framework from the jQuery CDN. Also notice in-keeping with tradition, we are loading the compressed jQuery and jQuery Mobile files so that we have the smallest possible footprint we can have:

```html
<meta name="viewport" content="width=device-width,minimum-
scale=1.0,maximum-scale=1.0,user-scrolable=no">
```

Now, you may or may not know this line, depending on how much mobile development experience you have. This line is important as it really makes a difference to how your website or application looks on a mobile browser. Apple started it for use on iOS Safari but has since been adopted by others. If you did not have this line in your project, the page we have would be rendered "zoomed out" basically. The user would have to double-click or zoom in via other means to see the page. By having this, we make sure mobile browsers are on their best behavior in helping your project look fabulous when the user visits:

```html
<div data-role="page">
```

This is the start of jQuery Mobile specific code. For the most part, our project is straight HTML5. There is very little actual JavaScript involved, but we have these weird div lines sprinkled throughout. These are jQuery Mobile tags. It uses the standard div tag but adds on these things called data roles. In this instance, we are giving div the role of a page. The page role is the main role in jQuery Mobile. We'll discuss this in more detail in another chapter. For now, let's keep moving.

```html
<div data-role="header">
<h3>Web Development Books</h3>
</div>
```

Here we are using another role, the header role. This allows us to display "Web Development Books" in a different manner than the rest of the page. As you might have noticed in the preceding screenshots, the entire text in the header is not visible. Whenever the text in the header is long, it will be truncated and appended by the ellipses (...). So make sure that you do not have really long names in the header. However, if long names cannot be averted, you can override the jQuery Mobile CSS to make sure that the text in the header is not truncated.

```
<div role="main" class="ui-content">
<ul data-role="listview" data-inset="true" data-divider-theme="b">
    <li data-role="list-divider">JavaScript</li>
    <li>
      <a href="#">Object Oriented JavaScript</a>
    </li>
    <li>
      <a href="#">JavaScript Unit Testing</a>
    </li>
    <li data-role="list-divider">PHP</li>
    <li>
      <a href="#">RESTful PHP Web Services</a>
    </li>
    <li>
      <a href="#">Instant RESS</a>
    </li>
  </ul>
</div>
```

This is the heart of our page, the content. Within the `div` tags, we are creating an unordered list, which jQuery Mobile then renders for us as buttons almost. If we had the links going to existing pages rather than the dummy link, the user could click the list items and navigate through the project. Again, we will be discussing these in much greater detail in another chapter.

Are you still with us? Everything might seem a bit overwhelming right now, even though we hope it doesn't, but as we go and you write more code, things will get easier and you'll see how powerful jQuery Mobile is without having to write a lot of complicated code.

Summary

In this chapter, we set up a local development server using XAMPP, and we installed an IDE to use for the book with Aptana Studio 3. We looked at the different ways you can download jQuery Mobile for local use in your project and then you saw how we can use it from the different CDNs, along with the advantages and disadvantages of using a CDN. Once all the basic setup was completed, we created our first jQuery Mobile project.

In the chapter that follows, we will take a look at the different tools and techniques that will help us not just in the development process, but also help us test our application as we develop it.

2
Tools and Testing

We started looking at the framework in the previous chapter and in this chapter we are going to look at some tools that we can not only use in the development of our applications, but in the design and testing of them as well.

No matter how good a framework is in terms of the features it offers, it isn't any good without proper tools to go along with it to write the next great app. Further, you also need good testing tools to test the application you created using this framework. Nobody, no matter how much of a rock star they may be, can write perfect code the first time. Unless of course you're John Carmack—we're convinced that the code that man needs is just willed into existence by his mind, and it's perfect the first time, every time.

Having said that, shall we begin our journey to explore some tools that we can use for our development and testing?

Overview

There are many tools you can use that will make your life easier in mobile web development, many more than we could ever cover in this book. We will first look at a wonderful tool that will make your life easier when it comes to styling your jQuery Mobile applications: **jQuery ThemeRoller**. Next up we will take a more in-depth look at the IDE Aptana Studio and how we can begin to harness the power of this powerful, yet free, application. Finally, we will look at **Screenfly** and how we can use it to test the responsiveness of our mobile web applications and how the site can look and behave on different devices and browsers.

Theming with jQuery ThemeRoller

jQuery Mobile features a very robust and powerful theming system. You can easily control the look of the application down to the most miniscule of details. The framework even allows you to have various color swatches that you can switch between on-the-fly as well. Before we look at ThemeRoller, maybe we should talk briefly about jQuery themes.

The basics

When you download jQuery Mobile, you get five color swatches, A–E. These swatches were created with a great deal of attention paid to readability and usability, and range from high-contrast (swatch A) to the lowest contrast (swatch E). In addition to these differing contrasts, the themes are also arranged so that the most prominent areas of the page, such as the headers and buttons, get more contrast compared to areas such as the footer. When you create your own theme, you should adopt this model of thinking as well.

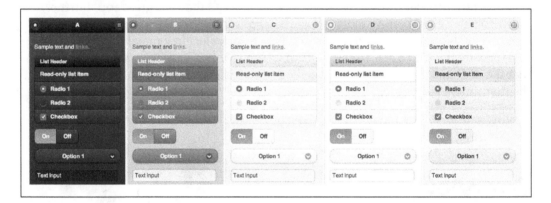

 For more information on jQuery theming, we highly recommended checking out one of the many books on the subject published by Packt Publishing.

Rolling into your own theme

Now that we understand what a theme is, let's look into a very powerful theming tool rolled out by jQuery Mobile—ThemeRoller. ThemeRoller is a web-based application that started as an extension for jQuery UI. Through this web application, you can create a theme with three swatches in a matter of minutes. Writing CSS code is not required; everything can be created through the use of point-and-click and dragging. While you are developing the theme, you can also use the real-time preview functionality to immediately see your creation in action. ThemeRoller also integrates in Adobe's Kuler product that allows you to see some premade color selections and add them to your current ThemeRoller color selection.

To get started, let's go to the ThemeRoller website at: `http://themeroller.jquerymobile.com`. Once you get there, you'll see a screen similar to the following screenshot:

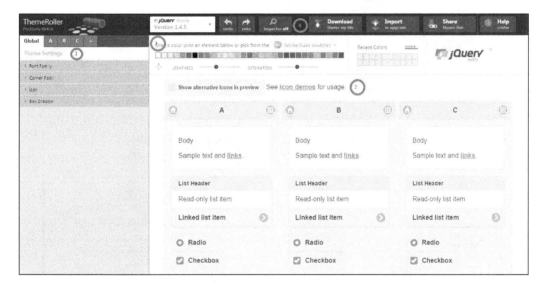

The screen can be broken up into four parts:

- Inspector
- Preview
- Color
- Tools

Inspector

This is the core of the ThemeRoller system. Here is where the magic happens, and you create your theme. The **Global** tab, which is the default tab, is where you control the font of your theme, how rounded the corners are, and the box shadow, and even change the icons. These options are universal across all the swatches in the current theme. The **Active State** accordion will also give you a current global snapshot of all the swatches as you make changes via the **Color** menu, as shown in the following screenshot:

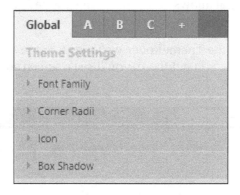

The three lettered tabs going across the top (**A**, **B**, **C**) with a **+** sign, are your color swatches. Here you can control the color of every aspect of the theme, from the header bar to the buttons. You may create up to 26 swatches for one theme.

One very important piece of information. By default, you have the three swatches. You are required to have three swatches in a theme, even if you only need one. So, if you decide that you just need one swatch, you will need to duplicate the swatch two times (or simply leave the other two swatches alone) to satisfy this requirement. If you create three new swatches, the theme that will be downloaded will have the CSS for all these 3 swatches.

To make a new swatch outside **A**, **B**, or **C**, simply click on the **+** sign or the **Add Swatch** button at the bottom of the **Preview** area.

Preview

The Preview section of ThemeRoller is where you will see the theme take shape in real-time, as you make your changes in the **Inspector** area. You can see three areas that currently all look the same. These three panels represent your swatches. Since by default we have three swatches, we have three panels here. When you add another swatch, the **Preview** area will add another screen to represent the new swatch:

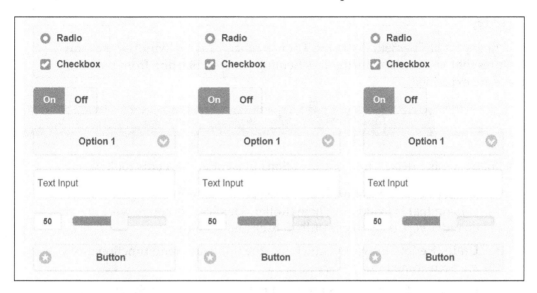

These panels show several of jQuery Mobile's components, so you can see how your color and style choices affect the vast majority of the framework, and can fine-tune the theme to your liking.

Color

Here you will find several color options along with 20 empty boxes. These empty boxes hold the last twenty colors you have used in this theme. Underneath the color choices, you will see two sliders: **LIGHTNESS** and **SATURATION**, as shown in the following screenshot. Lightness controls the light and dark tones of the on-screen color options, while saturation controls how vibrant the color is. To read more about Adobe Kuler – also called as Adobe Color CC – visit https://color.adobe.com/create/color-wheel/:

One of the things that make this tool so easy to use is while you can use the Inspector area to add in colors via hex codes and the like, you can actually drag a color from this toolbar and drop it onto the controls in the Preview area, and I mean *any* control. If you want to change the background of a button, drag the color down and drop it on the button. And if you want to change the background of your lists, just drag-and-drop. This tool is that awesome!

Tools

The final part of ThemeRoller is the **Tools** toolbar. This toolbar offers several features that you can use during the theming process. Starting from the left, these are explained:

The following is a description of the options present in the **Tools** toolbar:

- **jQuery Mobile Version 1.4.2**: Here you can change the jQuery Mobile version that is loaded in ThemeRoller. This way, if you're working on an older version, you can still go back and theme it.

- **Undo/Redo**: This works much like the undo and redo functionality you are accustomed to.

- **Inspector off**: This inspector is not to be confused with the Inspector component of this tool. This inspector, when turned to on, will allow you to inspect any of the widgets in the Preview area. When you inspect an item by clicking on it, the widget will highlight with a blue outline and the corresponding entry in Inspector will expand so you can see the properties of it.

- **Download**: This allows you to download the theme you are currently creating.

- **Import**: This button will let you import a previously created theme to edit.

- **Share**: Here you can get a generated URL that you may send out to folks to see what you've done so far. This is a great asset when others need to approve your theme or you need some help with it.

Creating a theme

Now that we've looked at what this program can do, let's put it to use and create a theme that we will apply to the project we created in the previous chapter.

When you start to create a theme, you have two paths you can take. You can create from scratch and just start making changes after the page loads up or you can import a previously created theme. Since it is fairly straightforward when editing an imported theme, we will look at creating one from scratch.

In this method, we'll simply start changing colors and options to our heart's content. To get started, let's make some global changes. Make the following edits in the **Global** tab of the Inspector:

- Font Family:
    - **Font:** `arial,sans-serif`

- Corner Radii:
    - **Group:** `0.3em`
    - **Buttons:** `1.5em`

- Box Shadow:
    - **Opacity:** `45`
    - **Size:** `5px`

Let's discuss these changes briefly. We are making `arial` the default font, and we're making the border of the groups a little more crisp but making the buttons a lot rounder. Then finally we are making our elements have a greater shadow, giving them a 3D illusion.

Now we are going to add some color to swatch A. Let's start first by changing the background color and dragging the tan color, the fifth color over from the right on our color selector (#c7b299), and dropping it on the background of swatch A. Now drag the blue, 12th from the right (#29abe2) and drop it on the body element of the swatch. This will also change a few of the other elements that are linked to this color as well. That's fine. Now drag the lighter blue that is just to the left of the blue we just picked (13th from the right and #33ccff) and drop it on the buttons. Again you'll see some additional items change color. So far, your swatch A should look as shown in the following screenshot:

Those blue links are a bit hard to read on that blue background, so let's change that. Click on the swatch A and expand the **Link** accordion. You'll see **Link Color** with a **+** sign off to the right. If you click the **+** sign, you change the color for hover, active, and visited links. For now, let's change **Link Color** to #f0eb94. Now our screen swatch should look as shown in the following screenshot:

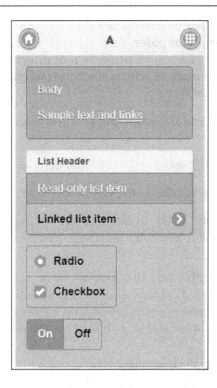

So we now have a theme with a light brown or tannish background (hex value #c7b299) with radio, checkbox, and active buttons that have a light blue background (hex value #33ccff) with black text (hex value #000000). For the body, we have another blue background (hex value #29abe3), white (hex value #ffffff) text, and yellow links (hex value #f0eb94). If you wish, you can see the theme we just created; visit http://themeroller.jquerymobile.com/?ver=1.4.2&style_id=20140417-38.

Now that we have some changes done, let's go ahead and download the theme and load it into our project.

Click on **Download** and you will be prompted with a screen that tells you how to use this theme in your project, as well as a prompt to name the theme. When we download the theme, we will be given two CSS files: a compressed one and an uncompressed one, and these files are to be used in the same fashion that the regular jQuery Mobile CSS files are used. The compressed minified (.min) version is recommended for production use and the uncompressed file is to be used for development.

After you have named the theme (we named ours `Chapter2`), click on **Download** and save the ZIP file to your computer. After it downloads, unzip it and copy the themes folder over to your `chapter 1` project's root directory. Now open up Aptana Studio 3 and if you haven't worked with it since the last chapter, it should load your previous chapter automatically.

Open up `index.html` and add the following line just before the main jQuery Mobile CSS file:

```
<link rel="stylesheet" href="themes/Chapter2.min.css" />
```

Your code should now look similar to the following code:

```
<link rel="stylesheet" href="themes/Chapter2.min.css" />
<link rel="stylesheet" href="http//code.jquery.com/mobile/1.4.5/
jquery.mobile-1.4.5.min.css" />
```

Look down the code to roughly line 18 where we have `<ul data-role="listview" data-insert="true" data-dividerthem="b">` and change it to `<ul data-role="listview" data-insert="true">` so that we remove the B swatch. Load up your project in your browser and you should see the new theme in action.

> **Theming, the traditional way**
>
> When you download jQuery Mobile, you get the CSS for the included theme. If you are so inclined, you can load up all the CSS files and hack away until your dream theme is created; the use of ThemeRoller is not required!

Using Aptana Studio 3

Aptana Studio 3 is a very powerful open source IDE. It is built from Eclipse and supports JavaScript, HTML, DOM, CSS, PHP, Ruby on Rails, and more. In addition to these supported languages and technologies, Aptana Studio also supports Git, deployment through FTP/SFTP/Engine Yard and other setups, a strong debugger with all the debugging goodness you'd expect, code assist, and finally a built-in terminal.

In *Chapter 1, Getting Started*, we installed it as a standalone program; however, if you are an avid Eclipse user, you can also install it as a plugin. Since we will be using this software for the remainder of the book, we should spend some time getting familiar with it.

The interface

We'll start by looking at the various aspects of the Aptana Studio interface. Due to scope constraints, we will only look at the pieces we will be using the most. For a more detailed look at Aptana Studio, check out *Aptana Studio Beginner's Guide, Thomas Deuling, Packt Publishing*.

Toolbar

Here, we have a lot of the standard icons and commands you'd expect to see in an IDE. Starting left to right, the toolbar will be explained:

The following is a description of the options:

- The new icon creates a new project, file, folder, and more.
- The save icon saves the current open file.
- The save all icon saves all the currently loaded files.
- The print icon prints the current file.
- The open URL icon will load the source code of the specified URL.
- The toggle breakpoint icon creates a breakpoint at the current location of the cursor.
- The debug icon launches the current project in Firefox (default) and changes other debug properties. Browsers in this option will need the respective debugging tool installed.
- The run icon is the same as debug icon but does not include the debug information.
- The open a PHP type icon allows you to inspect PHP elements.
- The next four icons toggle on/off certain windows in the IDE.
- The terminal icon opens a terminal window.
- The themes icon allows you to change the current look of the IDE.

Project Explorer

This window will list all the files and folders associated with the current project. You can right-click in this area and import/export files, create new files and folders, and create a new file from a specified template. This window also supports the standard operations such as renaming and deleting files. The window looks like the following screenshot:

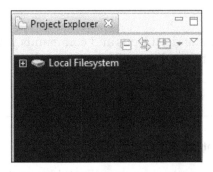

Console, terminal, and problems

This window will show various output messages during the debugging process, compilation errors, and, when you are using the **Terminal** tool, this will be the actual window that is used:

The Editor window

The heart of Aptana Studio, the editor. This screen is where all the magic happens. The editor has code completion, syntax highlighting, and code folding. The window looks similar to the following screenshot:

```
 1   <!DOCTYPE HTML PUBLIC "-//W3C//DTD HTML 4.01 Transitional//EN"
 2   "http://www.w3.org/TR/html4/loose.dtd">
 3   <html xmlns="http://www.w3.org/1999/xhtml">
 4       <head>
 5           <meta http-equiv="Content-Type" content="text/html; charset=utf-8" />
 6           <title>New Web Project</title>
 7       </head>
 8       <body>
 9           <h1>New Web Project Page</h1>
10       </body>
11   </html>
12
```

Suggested customizations

The default IDE theme is Aptana Studio. Personally for us, it's a bit rough on the eyes so we always change the theme to Dreamweaver. It's a very clean theme that is a lot easier on the eyes. Of course this is a personal preference and you may like the default, but take a few minutes and look through each theme to see if there is one you like better. The more comfortable the IDE is on your eyes, the better the coding experience will be for you.

We need to specify our XAMPP instance so that it can be used when we execute the project. Technically, if you are storing all of your Aptana projects in the XAMPP htdocs folder, this wouldn't be used, but just to be safe or if you have a web server you'd rather use, this will come in handy.

1. Go to the **Project** menu and choose **Properties**.
2. Click on **Preview Settings** and then click **New**.
3. From this dialog box, choose **External Web Server**.
4. Enter any name you wish in the **Name** field.
5. For the **Base URL** option, put in http://localhost (if you had to use a different port number from 80, be sure to specify that as well).
6. **Document Root** should be <XAMPP installation directory>/htdocs.
7. Click **OK** twice and the server is set for this project.

If you use GitHub or have a JIRA instance, you can set the username and password for these services under the **Preferences | Accounts** window, which can be found under the **Window** menu.

Navigate to **Preferences | Editors**, you can set the number of lines in a tab for each of the supported languages. I use the default but I know some people prefer to increase or decrease that number, so this is where you can make that change.

Creating a new project

Aptana Studio supports four different project types:

- **Web**: This is your standard web project that allows you to create an empty project or create one based on the HTML5 Boilerplate.

- **PHP**: This is recommended for projects that will be using PHP. You can also set the PHP release compatibility for the project.

- **Ruby**: Here you can create a new Ruby project based on the Ruble template or an empty project.

- **Rails**: The last project is a Rails project.

For the majority of this book, we will be using either Web or PHP.

1. Let's go ahead and create a web project based on HTML5 Boilerplate.
2. Navigate to **File | New | Web Project**.
3. Select the **HTML5 Boilerplate** template.
4. Click **Next**.
5. Name your project anything you wish; we're using `aptanatest`.
6. Make sure the default location checkbox is checked and the location is pointing to your XAMPP `htdocs` folder.
7. Click **Finish**.

Your **Project Explorer** screen should look similar to the following screenshot:

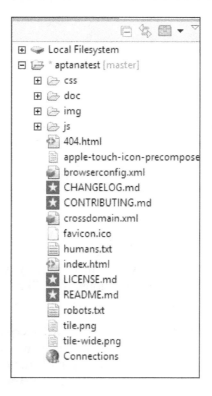

All of these files are part of the extremely popular HTML5 Boilerplate template. You can see more about it at `https://github.com/h5bp`.

Let's see what this bad boy looks like when we execute it. Click on **Run**. Anything happen? Were you one of the lucky ones that knew what to do to get Aptana Studio to execute the code? There is one nuance of Aptana Studio that bothers us and gets us about 9 out of 10 times when we go to run a project. If you do not have the main file of the project, in this case `index.html`, selected when you click **Run**, nothing happens. No error, no dialog box asking you to select a file, nothing. We admit; we're not sure if this is something to do with Eclipse itself, or whether it's just Aptana Studio. So if your project did nothing when you clicked on **Run**, select `index.html` and click **Run** again.

This time around you should have seen something similar to the following screenshot:

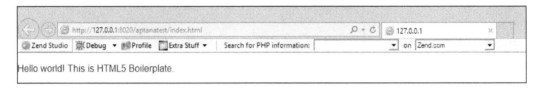

At first glance, you may be disappointed that all those files just give this one line of text. HTML5 Boilerplate is a great start to writing HTML5 applications. It puts in a lot of the gotchas you may come across so that you don't have to worry about them. So while this sample is very miniscule, it is an extremely powerful and complex tool in the backend.

Testing with Screenfly

Screenfly is a great tool that can assist you with seeing how your website or application will look on a variety of devices and resolutions. Also, like every other tool in this section, it is completely free to use!

Open up your browser and go to `http://quirktools.com/screenfly/`. You will be presented with a very simplistic and easy-to-use website. Don't let this simplicity fool you; once you submit a website you want to look at, you'll see the true power of Screenfly. You'll see a screen similar to the following screenshot:

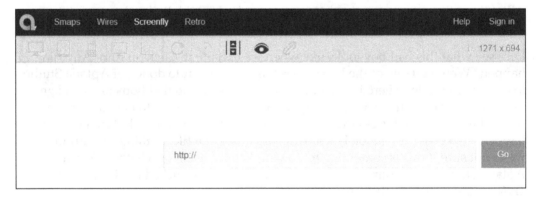

Getting started with this tool is extremely easy. Simply input the URL of the website or application you would like to look at; you can even use the localhost address of our project from the previous chapter.

Once you hit **Go**, you will see your website in the middle of the screen and it will appear as shown in the following screenshot:

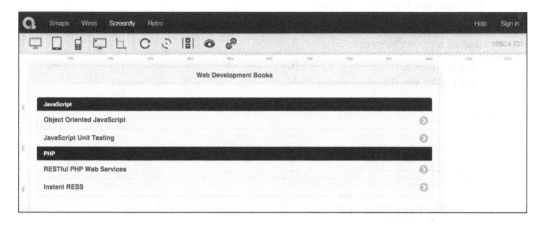

Now you will have several icons active at the top of the page. These icons represent various devices and options. You can also access each option through a keyboard shortcut; simply press the letter in parentheses after the icon name to access its options.

The interface

Starting from the left, you have the following icons:

- **Desktop** (shortcut key: *D*): This icon allows you to view the site on various desktop resolutions.

- **Tablet** (shortcut Key: *T*): From here you can see how your page appears on devices such as the iPad, iPad mini, Galaxy Tab, and others.

- **Mobile** (shortcut Key: *M*): This menu has various phone handsets such as the iPhone 3/4/5, RAZR V3/V8, Galaxy S3/4, and more.

- **Television** (shortcut Key: *V*): Here you can see how your website looks on HDTV modes such as 720p and 1080p. This could be great for setting up websites or web applications that will be running in kiosks.

- **Custom Screen Size** (shortcut Key: *C*): If you need to check how it would look at a resolution that is not covered by any other options, here is where you can do it. Simply enter in the resolution and click **Apply**.

- **Refresh** (shortcut Key: *F*): This allows you to refresh the page that is loaded. This comes in handy when looking at new changes you've made.

- **Rotate Screen** (shortcut Key: *R*): Pushing this icon will simulate a screen rotation so you can see how your project will respond.

- **Allow Scrolling** (shortcut Key: *S*): By default you cannot scroll on the page that Screenfly loads, but by clicking this icon and turning scrolling on, you can see the full website.

- **Use Proxy Server** (shortcut Key: *P*): This accesses your project through the use of a proxy server.

- **Share** (no shortcut key): This allows you share this session out to anyone you'd like via the generated link.

The icons look as shown in the following screenshot:

More information

Go ahead and spend some time with this tool. Try out as many different websites and devices as you can. You may be surprised how one of your websites looks when your users try to visit it from their iPhone.

One quick note about using this great tool. While you can test the way your site looks on various devices, Screenfly does not give you any indication of how it may act on the selected devices. If you use functionality such as offline storage, geolocation, or anything else that uses any device functionality, I implore you to still test them on actual devices. Do not rely solely on Screenfly being your only means of device testing.

While you are on this page, you may notice the other links at the top of the page such as **Wires** or **Retro**. These are some other cool tools that you can use in any web development scenario. Retro shows you some important information about your current browser (the one you are accessing Screenfly with) and Wires lets you create wireframe UI prototypes.

Summary

In this chapter, we talked about several tools that will help us on our path to jQuery Mobile development enlightenment.

We saw how easy it is to customize our jQuery Mobile applications with ThemeRoller. We were able to create a theme without touching any CSS code and saw how we can apply that theme to our current project.

You then learned more about Aptana Studio, the IDE of choice for this book, and saw the power of a free IDE that rivals the paid developing software packages. We covered a few recommended customizations and changes that will make developing easier. And finally, while looking at Aptana Studio, we touched briefly on the HTML5 Boilerplate.

Finally, you saw a tool that will allow us to test our applications on a variety of devices to see how they would look on other devices we may not have access to and talked about the importance of testing on the actual device whenever possible.

We will take a look at mobile design and how to make designs look and run better on any mobile device in the next chapter. We will explore feature detection and device detection in depth.

In the chapter that follows, we will look at different techniques that are helpful for developing for the mobile devices. We will look at the responsive web design techniques including media queries, feature and device detection, and much more. We will also be demonstrating how responsive web design can be combined with server-side logic. So see you in the next chapter!

3
Mobile Design

Before we go too heavily into jQuery Mobile itself and how to use it, we should probably step back and look at mobile design and how to get your web applications to look good and behave well on any mobile device your user may have.

Overview

Responsive web design (**RWD**) is a web design methodology that allows you to design your websites and web applications so that they react well to various resolutions and screen sizes of whichever Internet-enabled device your work may be displayed on. In this chapter, we will cover:

- Some RWD techniques and practices
- Take RWD to the next level via **Responsive Design + Server-Side Components** (**RESS**)
- Explore libraries such as Modernizr and WURFL to see how these can enhance your application's user experience even further

Responsive web design techniques

To get started, let's take a deeper look at RWD and why it is important. By using a RWD approach in your applications and websites, you'll be able to ensure that, when users come to your site with their 55-inch, 1,080-pixel Internet-enabled television, they'll have the same experience as if they were visiting it from their iPad or Galaxy S5 phone.

You may be asking yourself why you should care if they're able to have the same experience on all the devices, or how likely will they be using their television to surf the Internet. The answer is, you should care as it's extremely likely as these days that users have a number of Internet-powered devices in their possession. They may be watching their favorite television program and then realize they need to check something on your website or access your web application, and they want that instant gratification. That is just the nature of our society now. We are always connected and we always want our information now, at this very moment, and for it to be useable and respond well to our television, tablet, and handheld gaming device, whatever it may be.

Are you scared yet? It may seem like you have a lot of work to do to ensure your next big project behaves and looks good on everything that accesses it but, to be honest, it is not as hard as you might think, especially if you take this approach at the beginning of your project's lifecycle.

RWD is made up of a series of methods that are not overly difficult to execute. These methods include flexible grids and layouts, careful image use, and CSS media queries. By using these principles, you'll see in no time that your website can handle any screen size, resolution, orientation, and device, no matter what the user throws at it.

Flexible layouts

Flexible layout or flexible grids (whatever you prefer to call it) are one of the critical pieces of RWD. What do we mean by flexible layouts? With flexible layouts, CSS completely drives the positioning of items on the page, your margins, and spacing. This CSS is of course used in conjunction with media queries to make sure the page reacts properly to the device it is being viewed on.

You may be saying to yourself that of course CSS is used for all that, but as always in life there is a caveat. You can no longer use pixels for your CSS values. With today's devices, not all pixels are equal; it is no longer a 1 for 1 pixel. On device A, 1 pixel may truly be 1 pixel. However on device B, 1 pixel could actually equal 5 pixels on device A. Still with me? To overcome this, in RWD we use percentages or **em**. If you're not sure what an em is, it is a unit of measurement based on a font's point size. Percentages and em allow you to do relative sizes instead of absolute size with pixels.

Images

Image handling can play a huge role in your RWD implementation. Unfortunately, it is something of a wildcard with RWD. There are several ways you can handle images. One of the most popular uses is with CSS and its *max-width* property. For example:

```
img { max-width: 100%; }
```

This will then keep the image at 100 percent of the browser screen width and will scale down with the screen width. With this method, you scale the image with CSS rather than declaring the height and width via HTML. Of course there is a catch with this. Some versions of Internet Explorer do not fully support the max-width property of CSS, so in your IE-specific stylesheets, you will need to use `width: 100%` instead.

There are also some JavaScript tricks you can use to make your images responsive, however one thing you must consider is that image resolution and their download times are crucial. If you load a 960-pixel-wide image onto a mobile site where the resolution of the website is 320 pixels, the initial 960 pixel image is going to be loaded and then scaled down. There will be a performance hit, not to mention that there is a chance the image could be horribly messed up from the scaling. Later in this chapter, we will look at another method you can use to effectively handle images.

Media queries

Media queries were introduced in CSS 2.1. These media types were `screen`, `print`, and `handheld`. CSS3 drastically expanded on these by introducing types such as `max-width`, `device-width`, `orientation`, and `color`. Any device introduced after the introduction of CSS3 will support these types. This is a good thing for you as you can definitely use media queries to target CSS stylesheets to these devices and, if the browser does not support them, they will be ignored.

Media queries in action

Let's take a quick look at an example use of media queries:

1. Open up Aptana Studio and create a new project by clicking on **Web Project**.
2. Add two files to the project, one named `mobile.css` and the other one named `desktop.css`.
3. Open up `mobile.css` and add the following code:

```
#content {
width: 95%;
margin: 0em 0em 1em 0em;
```

```
background-color: red;
color: white;
}
```

4. Now open up `desktop.css` and type this code:

```
#content {
width: 95%;
margin: 0em 0em 1em 0em;
background-color: black;
color:white;
}
```

The code that we just typed is standard CSS code. What we are doing is setting the content area of our page to be 95% of the screen with a 1-em bottom margin. We are then setting the background color of our content area to be either red (for mobile devices) or black (for desktop devices) with white text.

Now that we have our CSS and seen what it does, it's time to put it into action with a media query:

1. Open up `index.html`, which was created by Aptana Studio, and type in the following code:

```
<!DOCTYPE html>
<html>
  <head>
    <meta name="viewport" content="width=device-width,minimum-scale=1.0,maximum-scale=1.0,user-scrolable=no">
    <title>RWD Media Query Test</title>
      <link type="text/css" rel="stylesheet" media="all and (min-width: 320px)" href="mobile.css" />
      <link type="text/css" rel="stylesheet" media="all and (min-width: 1024px)" href="desktop.css" />
  </head>
  <body>
    <div id="content">
        I just did a media query!
    </div>
  </body>
</html>
```

2. Now load up this page in your favorite browser and you should see the following image:

3. You should see **I just did a media query!** with a black bar behind it. Now if you resize your browser to a smaller size resembling a mobile device, you should see something similar to the following screenshot:

4. Again you should see **I just did a media query!** with a red background this time. Now that we've seen the code in action, let's break it down and see what we just did.

The following code snippet is just your standard opening code on an HTML page:

```
<!DOCTYPE html>
<html>
<head>
```

The next part—the `meta` tag is of the utmost importance in mobile web development. Proper use of this tag ensures that the web page is displayed perfectly on your mobile browser:

```
<meta name="viewport" content="width=device-width,minimum-scale=1.0,maximum-scale=1.0,user-scrolable=no">
```

Here, `width=device-width` helps set the width of the viewport to the pixel width of the device in use. This prevents the user from resizing and dragging the page around. To disable the user from zooming into the web page, we make use of the user-scalable property. Also, `minimum-scale=1.0` and `maximum-scale=1.0` make sure that the web page is displayed at 100 percent size on the device screen.

Next up is the most important part of this exercise, the CSS media query. For the first line, we are saying for any device with a minimum device width of 320 pixels, load our `mobile.css` stylesheet. The next line tells the browser that, if the device has a minimum device width of 1,024 pixels, load our `desktop.css` file:

```
<link type="text/css" rel="stylesheet" media="all and (min-width:
320px)" href="mobile.css" />
    <link type="text/css" rel="stylesheet" media="all and (min-
width: 1024px)" href="desktop.css" />
```

In the preceding statements, you will notice the use of the keywords `all` and `min-width`. Let's talk a bit about these. When we use the keyword `all` and `min-width`, the media query will work on a desktop as well, just by resizing the browser window. This helps when the application is still in development and we need to test how the media queries behave. However, if you wish that the media query code should be restricted for mobile devices alone, you need to replace the keywords `all` by `only screen` and `min-width` by `min-device-width`. We encourage you to discover other properties that the media queries can take and how they behave differently.

The next bit of code is our page body, wrapped in the content CSS `div` we specified in our CSS stylesheets:

```
<body>
    <div id="content">
        I just did a media query!
    </div>
</body>
```

That wasn't so bad, was it? You can see how deep you can go with these media queries by loading specific CSS stylesheets for a wide variety of device screen sizes.

Of course, as always with great power comes great responsibility. Each media query can add a good bit of overhead to your application or website. Remember CSS is executed on the client side, so that means the page loads and then begins executing the media queries until a match is found. This can seriously impact the loading speed of your project, especially if someone is accessing your application from a slower cell connection (remember in some parts of the United States and the world, cell users still have pre-3G speeds) so while it would be tempting to do a media query for every `min-width` and `max-width` combination out there, your users might suffer for it.

We just touched briefly on the media query types out there. There is a lot more you can use media queries to test orientation, aspect ratios, and so much more. For the complete list of media query types, check out the W3 website at `http://www.w3.org/TR/css3-mediaqueries/`.

There is a variety of device widths out there thanks to the vast number of mobile devices. You can find a great comprehensive list of device widths at `http://viewportsizes.com/`.

Feature detection with Modernizr

Not all browsers are created equal and neither is their support for many aspects of HTML5. How can we adequately account for that? A small JavaScript library named Modernizr, which can handle what's known as feature detection, comes to our rescue.

Modernizr is a powerful JavaScript library that will allow you to test the browser for specific features. If you want to make a very hip HTML5 web application and are worried about the current divide on the HTML5 standard implementation across the browsers, you should consider using this incredibly powerful library. With it, you can check to see whether the browser supports the canvas tag, for example, or whether it supports some of the new HTML5 form elements such as tel. You name it and more than likely Modernizr can test for it.

In this section, we'll use Modernizr to test a few different popular HTML5 elements that we may use in our Mobile jQuery project later in this book.

Getting started

First we will need to download the Modernizr library. To do so, complete the following steps:

1. Go to `http://modernizr.com/download/`.
2. Here we can build the options we need for our project. I recommend when you go to use this library in production, you select only the options you want to use and test for, however for the purpose of this book go ahead and click Toggle on each of the three main sections (**CSS3**, **HTML5**, and **Misc**). Do not worry about downloading any extra detects from **Extra**, **Extensibility**, and **Non-core detects**.
3. After we have selected the detects we want, click **Generate** and you should see the entire JavaScript code load into the textbox at the bottom. You will also now see a **Download** button.
4. Click on the **Download** button and save it to your computer.
5. Open up Aptana Studio and create a new project by clicking on **Web Project** and name it `modernizr`.
6. Go to the location of your project in Windows Explorer or Finder, whichever is appropriate for your operating system, and create a new folder named `js`.
7. Unzip the `Modernizr` zip file you just downloaded to this newly created folder.

We will use this project for the next couple of sections so just leave Aptana Studio open.

Testing for geolocation

In this section, we are going to check to see whether the browser supports the geolocation API of HTML5 and if so, make use of it:

1. Go back to Aptana Studio and add a new file (by right-clicking on the project) and name it `geolocation.html`.

2. Add the following code to that file:

```
<!DOCTYPE html>
<html>
<head>
    <title>Geolocation Test</title>
    <script src="js/modernizr.js"></script>
    <script>
        function locateUser()
    {
        if (Modernizr.geolocation)
        {
            alert("Your browser can use Geolocation");
            navigator.geolocation.getCurrentPositi
on,errorHandler, {enableHighAccuracy : true});
        }
        else
        {
            alert("Sorry your browser doesn't like Geolocation");
        }
    }
    function showCurrentLocation(position)
    {
        document.getElementById("location").innerHTML = "Your
Latitude: " + position.coords.latitude + "<br /><br />" + "Your
Longitude: " + position.coords.longitude;
    }
    function errorHandler (error)
    {
        alert("Error loading position.\n\nError code: " + error.code
+ "\n\nMessage: " + error.message);
    }
    </script>
</head>
<body>
<div id="location">
    <input type="button" value="Where am I?"
onclick="locateUser()" />
```

```
    </div>
    </body>
    </html>
```

3. After you get the code typed, open your browser (any device will work) and you will be greeted with the following button:

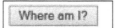

4. After you click the button, you will see a JavaScript alert that either says your browser supports geolocation or does not.

5. If your browser supports geolocation, after you hit **OK** to clear the initial alert you should see a popup or some other form of browser alert telling you that this website wants to use your current location. This is an important alert. Geolocation can only be used if the user allows it. If they click **Deny** when asked, your website cannot use the user's location on the site at all.

6. If you click **Allow**, you should see your latitude and longitude coordinates displayed on the screen as shown in the following screenshot (we've blacked out our coordinates):

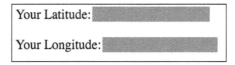

There is a lot of code there, so let's break it down.

In these first lines of the code, we are doing normal HTML header stuff. This line may be new:

```
<script src="js/modernizr.js"></script>
```

What we are doing here is loading up the Modernizr library:

```
function locateUser()
    {
      if (Modernizr.geolocation)
      {
        alert("Your browser can use Geolocation");
        navigator.geolocation.getCurrentPosition(showCurrentLocation,e
rrorHandler, {enableHighAccuracy : true});
      }
      else
```

```
        {
            alert("Sorry your browser doesn't like Geolocation");
        }
    }
```

In the preceding code, we are creating a JavaScript function that will first do a Modernizr check to see whether the browser supports geolocation. If the browser supports it, we let the user know via a JavaScript alert and we then assign `navigator.geolocation` to another JavaScript function named `showCurrentLocation` shown here:

```
function showCurrentLocation(position)
    {
        document.getElementById("location").innerHTML = "Your Latitude:
" + position.coords.latitude + "<br /><br />" + "Your Longitude: " +
position.coords.longitude;
    };
```

This code will look for the location `div` in our HTML code and then write the user's current latitude and longitude to the screen. This will only occur if the user gave permission for the website to view the location. To check this permission and handle it gracefully, we use the following code:

```
function errorHandler (error)
    {
        alert("Error loading position.\n\nError code: " + error.code +
"\n\nMessage: " + error.message);
    }
```

With this error handling code, we have three possible values that can be returned:

- `PositionError.PERMISSION_DENIED`: The user did not allow the website to access this information

- `PositionError.POSITION_UNAVAILABLE`: It was unable to determine position

- `PositionError.TIMEOUT`: The system timed out when determining the position

We're pretty sure you're probably thinking about some pretty good ways to use geolocation in your mobile web application. You could target ads to the user based on their location, serve them particular news, and more. Be careful with using it though. Some users do not and will not share their location with websites, especially if it is something they are not familiar with, so if your website requires use of it, you need to have a good reason to do so or you could lose a user.

Testing for web storage

Another great element of HTML5, and that we feel is one of the most powerful, yet extremely easy to implement, is web storage. Web storage is a handy feature of HTML5 that will allow you to store a small amount of information and can be a replacement for your typical session/cookie storage. You even have a sandboxed filesystem so you can write quite a bit of information to the web storage.

To take a look at web storage and how you can use it with Modernizr, let's look at another basic example. Instead of creating a whole new project in Aptana Studio, we'll expand on the geolocation example; this way we can see how we can blend multiple Modernizr checks in one project:

1. Open up the previous project we did in Aptana Studio if it's not open already.

2. The first bit of code we need is to add the following code to showCurrentLocation JavaScript function after document. getElementByID("location"):

   ```
   localStorage.setItem("prevLatitude", position.coords.latitude);
   localStorage.setItem("prevLongitude", position.coords.longitude);
   ```

3. The showCurrentLocation (position) should now look as follows:

   ```
   function showCurrentLocation(position)
   {
   document.getElementById("location").innerHTML = "Your Latitude: "
   + position.coords.latitude + "<br /><br />" + "Your Longitude: " +
   position.coords.longitude;
   localStorage.setItem("prevLatitude", position.coords.latitude);
   localStorage.setItem("prevLongitude", position.coords.longitude);
   }
   ```

4. Now add the following below the closing of our location div:

   ```
   <p>
   My Previous Coordinates: <br />
   <textarea id="coordhistory" width="300" height="300"></textarea>
   <script>
   if (Modernizr.localstorage)
   {
   alert ("Web storage capable");
   var savedpositions = document.getElementById("coordhistory");
   savedpositions.value = "Previous Latitude: " + localStorage.
   getItem("prevLatitude") + " Previous Longitude: " + localStorage.
   getItem("prevLongitude");
   }
   ```

```
else
{
alert("Your browser does not support Web Storage");
}
</script>
</p>
```

5. Our `<body>` section of the page should now look as follows:

```
<body>
<div id="location">
    <input type="button" value="Where am I?"
onclick="locateUser()" />
</div>
<p>
My Previous Coordinates: <br />
<textarea id="coordhistory" width="300" height="300"></textarea>
<script>
if (Modernizr.localstorage)
{
alert ("Web storage capable");
var savedpositions = document.getElementById("coordhistory");
savedpositions.value = "Previous Latitude: " + localStorage.
getItem("prevLatitude") + " Previous Longitude: " + localStorage.
getItem("prevLongitude");
}
else
{
alert("Your browser does not support Web Storage");
}
</script>
</p>
</body>
```

6. Now fire up the page through whatever device/browser you wish to use and you should immediately see a JavaScript alert telling you whether or not your browser supports web storage. Click **OK** to dismiss the alert.

7. You should see a blank text area element with the familiar **Where am I?** button from the geolocation example.

8. Click **Where am I?** and you should see the geolocation prompt and your coordinates displayed to the screen just like before.

9. Now refresh the page and you should see your previous coordinates loaded in the text area as follows:

Now let's break down the new code we just added:

```
localStorage.setItem("prevLatitude", position.coords.latitude);
localStorage.setItem("prevLongitude", position.coords.longitude);
```

This is the first bit of code we added for this example. What we are doing here is creating/setting a new localStorage variable named prevLatitude and prevLongitude to the respected coordinates we get from the geolocation API through the use of setItem method of the localStorage object. As you can see, it is extremely easy to create these web storage variables.

```
My Previous Coordinates: <br />
<textarea id="coordhistory" width="300" height="300"></textarea>
```

Here we are creating a standard HTML text area control that will hold the values of our localStorage variables:

```
<script>
  if (Modernizr.localstorage)
  {
    alert ("Web storage capable");
    var savedpositions = document.getElementById("coordhistory");
    savedpositions.value = "Previous Latitude: " + localStorage.
getItem("prevLatitude") + " Previous Longitude: " + localStorage.
getItem("prevLongitude");
  }
  else
  {
    alert("Your browser does not support Web Storage");
  }
</script>
```

Again this is the crux of our example. We first perform a Modernizr detect to see whether the browser supports localstorage. If it does, we let the user know via a standard JavaScript alert and we then assign our textarea to a JavaScript variable. We access our localStorage variables via the getItem methods, concatenate them with some text, and assign this value to that of the textarea. Notice a very key item here. Modernizr uses localstorage to check for the localstorage capability, while JavaScript uses localStorage as the object name. If for some reason your code isn't working correctly, make sure you haven't mixed up the two as JavaScript is case-sensitive.

We told you web storage was easy to use! Also you should note that while web storage is part of HTML5 specification, it has grown so much that it is actually being split off into its own specification. This is exciting for us web developers as it can really give some very cool functionality to web storage, making it even better.

This discussion concludes our look at Modernizr for this chapter. In the next section, we will look at device detection.

It may seem pretty cool that you can store a lot of information in the user's browser. We've seen some people declare this is a way to end server-side database engines. We cry a little inside when we see that. While it is true that you can store a lot of information in the browser, please remember a couple of things.

By default, you are limited to 2.5–5 MB depending on the device and browser. There is a way you can ask for more space from the user, but since that actually counts as space on their device, some users may not be willing to give your website or application that additional space. Because of this, you cannot depend on them to give it to you and require your application to make use of that space that you may not get.

Finally, since the user has to authorize you to have additional space, this means they are in control of that space. They can clear it at any time they want to. Since this could happen at any moment, you don't want to store their order history or anything like that in web storage. Storing information such as website preferences, sessions, or cookies is okay as those can be easily added back and the user doesn't lose anything. Also, of course, we shouldn't have to say this but you never want to store any user information such as credit card numbers, social security numbers, and so on in there so that a hacker could use it in the event the user loses their mobile device.

Device detection with WURFL

In the previous section, we looked at the role Modernizr plays in mobile development using feature detection. Now we look at another piece of the puzzle: device detection. Why do we need device detection you say? While we can see what features the browser supports, it is also nice to see what actual device the user is using.

Let's look at an example on why we would want this. You have been working on a mobile website and, during development, you notice that it behaves differently on a Galaxy Tab than it does on an iPad. You bang your head on the wall trying to get one CSS file to serve both but nothing works, and you think to yourself how great it would be if you could have the Galaxy Tab use one CSS file and the iPad use another. Well device detection allows you to do just that.

For this book, we will be using the cloud version of **WURFL** from **ScientiaMobile**. Why's that? Well, over the years, ScientiaMobile has changed so that the downloaded version of the WURFL library can only be used in open source projects that fall under the AGPL.

For their cloud service, ScientiaMobile has several licensing account options. Luckily, they have a free version that we can use to show off the capabilities of the system and get you started down the path of device detection.

Getting started with the ScientiaMobile cloud

Let's start taking a look at this service and how we can use it in our mobile project:

1. Go to `http://scientiamobile.com/wurflCloud/gettingStarted`. This page will guide you through the account registration process and generating your API keys.

2. After your account is created, click on **My Account** and then, under the section that says **Cloud Subscriptions**, click on your account name.

3. Here is your control panel, so to speak. You can add what capabilities you want to check for, see how many of your monthly detections you have used, and more.

4. Click on **Download Client Code** link and then click **Download** under the PHP logo.

 You may notice here that ScientiaMobile has libraries for various languages. For this book, we are going to use PHP but feel free to come back and check out the library for one of your favorite languages.

5. Once you have downloaded the zip file, open it up and find the `src` file folder.

6. Drag this folder out to your desktop for now.

7. Create a new PHP project in Aptana Studio and name it `wurfl`.

8. Add the `src` folder you just put on your desktop into the directory that your project resides in.

Now that we've got our account set up and our code downloaded, let's take look at that problem we had earlier with needing a different CSS file depending on the tablet device.

Device detection example – tablet brand

Before we can get going with the code, we have one more item we must take care of in our ScentiaMobile account. We need to let the cloud know what capabilities we want to test for. With a free account, we are limited to two capabilities but, on a positive note, we can change those capabilities whenever we want, which we will do a few times throughout this book.

1. Go back to the dashboard from step 3 in the previous section. Here you will see a link that says **API Keys** and you need to copy the API key as you'll need it here in a minute.

2. From the **API Keys** page, find the link **Capabilities** and click on it.

3. You will see an extensive list of capabilities. Note that this is just one category. If you look towards the top of the page, you will see a drop-down menu named **Group** with `product_info` showing right now. This is the category we are going to use for now.

4. Drag the following capabilities over to the section named `Your Capabilities`:

   ° `is_tablet`

   ° `brand_name`

   ° `model_name`

5. Click **Save**.

Finally, we are ready to code so let's go back to our project in Aptana Studio.

Open up `index.php` (create it if you have been using the empty project) and add the following code:

```php
<?php
//Include the WURFL cloud client
//Edit this path to reflect your project structure
require_once 'src/autoload.php';

//Creates a configuration object
$config = new ScientiaMobile\WurflCloud\Config();

//Setup your API keys
$config->api_key = 'Your Key Goes Here';

//Create the cloud client
$client = new ScientiaMobile\WurflCloud\Client($config);

//Detect the device
$client->detectDevice();

//Use the capabilities
$tablet = $client->getDeviceCapability('is_tablet');
$brand = $client->getDeviceCapability('brand_name');
$model = $client->getDeviceCapability('model_name');
$os = $client->getDeviceCapability('device_os');
?>
<html>
    Tablet: <?php echo $tablet; ?> <br/>
    Brand: <?php echo $brand; ?> <br/>
    Model: <?php echo $model; ?><br/>
    OS: <?php echo $os; ?><br/>
</html>
```

Now load this page up on a tablet device if you have one handy. You could also use one of the Android or iOS simulators you may have.

 Unfortunately for this our Screenfly tool will not work as it will not send the appropriate information to the page that WURFL needs for the device detection.

On our iPad, we see this:

It was able to detect a tablet device, so it returned `true` or 1 (0 would be false). It detected Apple for the brand, iPad for the model, and of course it is running iOS.

Do you see the power of this library yet? Let's break the code down:

```
require_once 'src/autoload.php';
```

Here we are loading the WURFL library to use in our application.

```
$config = new ScientiaMobile\WurflCloud\Config();
```

Now we are instantiating a new instance of the `Config()` PHP object.

```
$config->api_key = 'Your Key Goes Here';
```

This is where you paste your API key into. Please note you do need to wrap it in quotes.

```
$client = new ScientiaMobile\WurflCloud\Client($config);
```

Here we create an instance of the `Client()` object and we pass it our `$config` object so that the library can use our API key.

```
$client->detectDevice();
```

This is the heart of library. It performs device detection and loads the capabilities we wanted, into memory.

```
$tablet = $client->getDeviceCapability('is_tablet');
$brand = $client->getDeviceCapability('brand_name');
$model = $client->getDeviceCapability('model_name');
$os = $client->getDeviceCapability('device_os');
```

Here we assign the result of each capability to a PHP variable. True, we can just use `$client->getDeviceCapability('is_tablet');` but, since we are going to be using it pretty often, let's assign it to a variable so we can easily access it and save some typing.

```
<html>
    Tablet: <?php echo $tablet; ?> <br/>
    Brand: <?php echo $brand; ?> <br/>
    Model: <?php echo $model; ?><br/>
    OS: <?php echo $os; ?><br/>
</html>
```

This is just a small piece of HTML to display what WURFL picked up for our device.

We can now see that the library is working properly. This is all we are going to do for this piece. In our next section, we'll expand on this project, add in some Modernizr, and see how the two can work together to make a web application.

The sheer volume of device information you can check for and retrieve with WURFL is phenomenal. You can dig through each of the drop-down menus to look at them, or you can refer to this handy list on their website that shows you all the information you can look for, what type of value it returns (Boolean, string, and many more), and a description of it. To view this list, visit `http://scientiamobile.com/wurflCapability`.

RESS introduction

So we've looked at responsive web design. It seems extremely powerful but a lot of it takes place on the client side. Wouldn't it be very advantageous to have some of that work take place server-side so that it will be more efficient for your users? Enter RESS.

It is an answer to the problems that plague standardized RWD. With a large number of media queries, coupled with scaling images on the client, your website can become quite large. The larger the website, the longer it takes to load. By using a server-side language, such as PHP in this book, we can still execute some RWD but have our server handle a greater amount of the load, thereby speeding up page loads. Of course we are not going to just use PHP by itself, we are going to also use what we just learned with Modernizr and WURFL to make our applications truly responsive.

RESS is a fairly new technique, with the phrase first coined by Luke Wroblewski in 2011. By using it, we can actually do more RWD by dynamically adding/removing content based on the device, all the while using one codebase across the big website and the mobile website (eliminating the need for a mobile website or URL address) and much more. We'll take a brief look at it and how you can use it to enhance your RWD development.

To accomplish RESS for this chapter, we will be combining RWD, Modernizr, WURFL, and PHP. We will take the project we started building in the last section with WURFL and expand it. So open it back up in Aptana Studio if it is not already.

Getting started with RESS

What we are going to do with this small project is detect whether the user is on a tablet device and then show them accessories based on what operating system they are running on the tablet. This will be a basic project yet it will showcase what you can do with RESS.

First we need to copy a few files over to this project. Copy over the CSS files from our Responsive Web Design project (rename `mobile.css` to `tablet.css`) and copy over the Modernizr JavaScript files from those projects.

Now you will need to create four new HTML files and add them to our project. Those files are listed as follows:

- `android.html`
- `apple.html`

- non_tablet.html
- other.html

Open up android.html and add the following code:

```
<iframe src="http://www.radioshack.com/family/index.
jsp?categoryId=2032381" width="85%""></iframe>
```

Next up, open up apple.html and add this code:

```
<iframe src="http://www.radioshack.com/family/index.
jsp?categoryId=3395029" width="85%"">
```

For other.html, add this:

```
<iframe src=http://www.radioshack.com/category/index.
jsp?categoryId=2032061" width="85%"></iframe>
```

What we are doing here is creating iframes within each of these pages that link to a specific page on Radio Shack's website. Each category is tailored to the appropriate Tablet OS.

Add this to non_tablet.hml:

```
<p>
Wait a second. It appears you do not have a tablet.<br />
I'm sorry you must have a tablet to view this page. Please check your
local Radio Shack for one and come back!
```

What we are doing here is simply telling the user that they must view this page on a tablet device in order to be able to use it.

Now open up index.php and replace all the contents with the following code:

```
<?php
//Include the WURFL cloud client
//Edit this path to reflect your project structure
require_once 'src/autoload.php';

//Creates a configuration object
$config = new ScientiaMobile\WurflCloud\Config();

//Setup your API keys
$config->api_key = '357280:gSusq6nAaYNoUGZm0xVcEvfi3421LwOy';

//Create the cloud client
$client = new ScientiaMobile\WurflCloud\Client($config);
```

```
//Detect the device
$client->detectDevice();

//Use the capabilities
$tablet = $client->getDeviceCapability('is_tablet');
$brand = $client->getDeviceCapability('brand_name');
$model = $client->getDeviceCapability('model_name');
$os = $client->getDeviceCapability('device_os');

if ($tablet)
{
    $stylesheet = 'tablet.css';
    if ($os == 'iOS')
    {
        $page_to_load = 'apple.html';
    }
    else if ($os == 'Android')
    {
        $page_to_load = 'tablet.html';
    }
    else
    {
        $page_to_load = 'other.html';
    }
}
else
{
    $stylesheet = 'desktop.css';
    $page_to_load = 'non_tablet.html';
}
?>
```

This is the PHP part of the code that should be followed by the HTML code, which just refers to the Modernizer library and then adds PHP references to the pages that need to be added.

```
<!DOCTYPE html>
<html>
<head>
    <title>RESS Test</title>
    <meta name="apple-itunes-app" content="app-id=544007664">
    <meta name="google-play-app" content="app-id=com.google.android.
youtube">
```

```
    <link type="text/css" rel="stylesheet" href="<?php echo
$stylesheet;?>" />
    <script src="js/modernizr.js"></script>
    <script>
    </script>
</head>
<body>
<div id="container">
    Below is some accessories for your <?php echo $brand; ?> tablet
device!
    <?php
        include($page_to_load);
    ?>
</div>
</body>
</html>
```

If we open this up on an iPad, we see the screen similar to the following screenshot:

We'll talk about that banner up top in a second. Now, if you view it on a desktop PC, you'll see the following:

> Below is some accessories for your generic web browser tablet device!
>
> Wait a second. It appears you do not have a tablet.
> I'm sorry you must have a tablet to view this page. Please check your local Radio Shack for one and come back!

Oh well, guess it's time to buy a tablet!

Actually, first it is time to break down that code we just wrote.

```
//Include the WURFL cloud client
//Edit this path to reflect your project structure
require_once 'src/autoload.php';

//Creates a configuration object
$config = new ScientiaMobile\WurflCloud\Config();

//Setup your API keys
$config->api_key = '357280:gSusq6nAaYNoUGZm0xVcEvfi3421LwOy';

//Create the cloud client
$client = new ScientiaMobile\WurflCloud\Client($config);

//Detect the device
$client->detectDevice();

//Use the capabilities
$tablet = $client->getDeviceCapability('is_tablet');
$brand = $client->getDeviceCapability('brand_name');
$model = $client->getDeviceCapability('model_name');
$os = $client->getDeviceCapability('device_os');
```

This is all the same code we used in our WURFL project. We're testing the same capabilities, so there was no point in changing the code or which detections we needed to do on the ScientiaMobile website.

```
if ($tablet)
{
    $stylesheet = 'tablet.css';
    if ($os == 'iOS')
    {
        $page_to_load = 'apple.html';
```

```
        }
        else if ($os == 'Android')
        {
            $page_to_load = 'tablet.html';
        }
        else
        {
            $page_to_load = 'other.html';
        }
    }
    else
    {
        $stylesheet = 'desktop.css';
        $page_to_load = 'non_tablet.html';
    }
```

Here we are doing a PHP if-else statement. We are first checking to see whether the user is on a tablet device. If the user is accessing the web page from a tablet device, we are assigning the `tablet.css` stylesheet to our PHP `$stylesheet` variable.

Continuing in the `if ($tablet)` check, we start seeing which operating system the user is running on their tablet. If the tablet is using an iOS, then we are assigning the `apple.html` file to our `$page_to_load` variable. If it is Android, we assign the `android.html` file. Of course if it is neither, then we load the `other.html` file that will be seen on Blackberry, Windows, WebOS, or whatever other tablet views the page and is not using Android or iOS.

Of course, if they are not using a tablet to view this page, we assign `desktop.css` to the `$stylesheet` variable and `non_tablet.html` to the `$page_to_load` variable.

```
<meta name="apple-itunes-app" content="app-id=544007664">
<meta name="google-play-app" content="app-id=com.google.android.
youtube">
```

We have a few standard HTML lines and then we come to the preceding meta tags here. What are those, you may be asking yourself? Let's say we had a kick-butt application based on our website in the Apple App Store and the Google Play Store. We could add these preceding lines, change the content values to the appropriate values that match our application in the respective store, and then, when users visit our page, they'd see the banners at the top of the page.

A quick note on these Smart Banners as they are called. For the Smart Banners to show up on Apple devices, the user must be using Safari on their iOS device. If they use Chrome or any other browser, they will not see the banner. The meta value is Safari-specific.

The same thing is true on the Android side of the house. The user must be using Chrome, not Opera or any other browser.

```
<link type="text/css" rel="stylesheet" href="<?php echo
$stylesheet;?>" />
```

Here we are loading our CSS stylesheet; however we are using PHP to dynamically load the appropriate stylesheet based on our detections from WURFL.

```
<div id="container">
    Below is some accessories for your <?php echo $brand; ?> tablet
device!
    <?php
        include($page_to_load);
    ?>
</div>
```

Finally, we have the content of our page. We dynamically load the file to the include directive of PHP to bring in whichever file we need to, again based on the information from WURFL.

So there we have it, a small working example of RESS. You may have noticed something; we did not do the first media query for mobile devices. In all honesty since we targeted tablet devices, the use of one was not warranted, as only tablet devices would see this page as we intended. However in a more real-world production scenario, you could do these checks and load in specific stylesheets based on Android or iOS devices. Then, within those stylesheets, you could do media queries to check the various screen sizes of Android tablets and a stylesheet that is targeted towards retina and non-retina iPads/iPad minis.

We know we said media queries could be hard on the client, but if you use them within PHP or another server-side language, you can control how they are executed. It is much better to run six media queries on an Android tablet, than run these six plus the four or five you have written for Apple devices.

RESS is extremely powerful in our opinion and we prefer it over standard RWD. Granted, both have their pros and cons, but we think you can be truly mobile-first with RESS.

Summary

In this chapter, we covered a lot, but sadly we barely scratched the surface of RWD and RESS. Packt has several books on RWD, so if you are interested in learning more, I highly recommend you check out one of them. If you are interested in a further look at RESS, you can check out *Instant RESS, Chip Lambert, Packt Publishing*.

In this chapter, we also looked at how you can use the JavaScript library to check the user's device browser for certain functionality. This allows you to not only tailor your websites and applications to be enhanced on certain devices, but also offer failovers in the event the visitor's device doesn't support a core feature you make extensive use of.

We then looked at device detection with ScientiaMobile's WURFL cloud PHP library. With it, you can check for specific devices themselves that your users may have. You can load the "Download my App from the Apple App Store" banner to your site and have it show on Apple devices. You can do the same with Android devices and Kindle Fires. The sky is the limit here.

To end the chapter, we then looked at how you can combine WURFL, Modernizr, PHP, and some RWD principles to make a website that goes beyond traditional RWD thinking and functions even better on the various devices that may load it.

These lessons, again while basic, will lay the framework for the rest of the book. We'll see how to combine these with jQuery Mobile and make one heck of a mobile application.

In the next chapter, we will begin building a customized mobile web application. We will take a look at some widgets and panels and see how we can make our mobile web application more powerful and interactive. See you on the other side!

In the next chapter, we will begin the development of our overall project—a web-based mobile application for a fictional Civic Center. Over the process of developing this application, we will explore a variety of jQuery Mobile widgets and events and a lot of other things that will help you learn in-depth about the jQuery Mobile platform.

4

Call to Action – Our Main Project

We are firm believers in the fact that the best way to learn something, especially a programming topic, is through hands-on, practical examples. So, this book will be no different. We are going to build a mobile web application for Anytown Civic Center. It will display events going on at the civic center, allow users to register for these events, and serve as a general mobile website for the civic center.

Overview

It is time to get down to business now. This project will display a list of upcoming events to the visitors, and allow them to register for the events that interest them. It will also have general information about the civic center on it, such as hours, driving directions, and more.

This project will be made using several features, or as they are known in jQuery Mobile: **roles** that make up the framework. These roles include pages, panels, popups, toolbars, and navbars.

As we go through this project, you may notice that jQuery Mobile uses the `<div>` element quite a bit instead of using, for example, HTML5's `<header>` and `<footer>` elements. This is to ensure backwards-compatibility with browsers that may not support HTML5.

Pages

Pages are the main unit within jQuery Mobile. The jQuery Mobile framework can support single pages or multiple internally-linked pages within one page (this may sound confusing right now, but you'll see shortly).

Pages themselves are made up of three components: header, content, and footer. Remember, as we mentioned previously, jQuery Mobile does not use the common HTML5 structure elements, so you will instead use these header, content, and footer implementations so that you can ensure the most cross-browser compatibility as you can.

Let's see an example of a page:

```
<div data-role="page">
  <div data-role="header">
    <!-- Header stuff -->
  </div><!-- /header -->

  <div role="main" class="ui-content">
    <!-- Content -->
  </div><!-- /content -->

  <div data-role="footer">
    <!-- Footer stuff -->
  </div><!-- /footer -->
</div><!-- /page -->
```

As you'll see throughout this chapter, whenever you start a new role, you use `<div data-role="">`.

One very important note: in order to make the most of jQuery Mobile, you must start the HTML page itself with the HTML5 doctype. Without it, you won't be able to take advantage of jQuery Mobile's power. Browsers that do not support HTML5 will simply ignore this doctype and continue working.

Let's go ahead and start working on our project. Create a new web project in Apatana Studio. I named mine `anytowncc`; you can use this name or choose another.

Now that the project is created, add a new a file to the project and name it `index.html` (if one was not already created for you by Aptana Studio).Add the following code to this file:

```
<!DOCTYPE html>
<html>
```

```
<head>
    <title>Anytown Civic Center</title>
    <meta name="viewport" content="width=device-width,
        minimum-scale=1.0,maximum-scale=1.0,
            user-scrolable=no">
    <link rel="stylesheet"
        href="http://code.jquery.com/mobile/1.4.5/
            jquery.mobile-1.4.5.min.css" />
    <script src="http://code.jquery.com/
        jquery-1.11.0.min.js"></script>
    <script
        src="http://code.jquery.com/mobile/1.4.5/
            jquery.mobile-1.4.5.min.js"></script>
</head>
<body>
    <div data-role="page">
        <div data-role="header">
            <h1>Welcome to Anytown Civic Center!</h1>
        </div><!-- /header -->

        <div class="ui-content" role="main">
            <p>Welcome to the website for Anytown Civic
                Center. We have a lot of great upcoming events
                    for you to check out.</p>
        </div><!-- /content -->

        <div data-role="footer">
            <h2>
                <i>1234 First Avenue Anytown, Anystate 12345
                    USA</i>
            </h2>
        </div>
    </div><!-- /page -->
</body>
</html>
```

We have seen all this code previously, so we will not break it down as we normally do. We declared our HTML5 doctype, we are loading in the jQuery and jQuery Mobile files from the jQuery CDN, and then we are creating our first jQuery Mobile page with a bit of placeholder text in it for right now.

One thing that we haven't used yet is data-theme="a". You could add a data-theme attribute on the page level to style the entire page, based on the colors of a particular swatch you want. You could use one of the six default swatches, or you could even make use of the custom swatch that we created using the jQuery Mobile ThemeRoller tool.

Now, you can open this page that we just created up in your normal web browser, or even better, you can see what it would be like on a mobile device via one you have laying around or use Screenfly. No matter which method you use, you should see something like this:

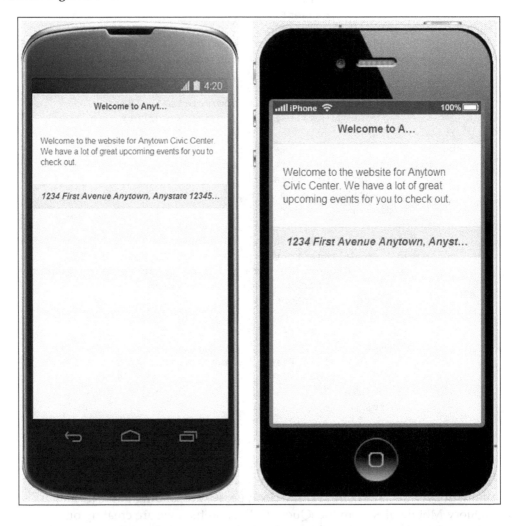

Now, we have the first page ready. We'll be adding more later on, but for now we are going to move on to panels.

Panels

jQuery Mobile has fantastic support for all kinds of different panels. You can use them to create menus, inspector panes, and so much more. You can have them attached to the left or right-hand side of the page and you have three options for them:

- `reveal`: This panel sits beneath the page and reveals as the page slides away
- `overlay`: This panel will appear on top of the page
- `push`: This panel and the page will be animated at the same time

When creating panels, they must be created within a page, as a sibling to the header, content, and footer components. The panel can be either before or after these elements, but not in between them. The source order does not really matter as long as the preceding constraints are met and as long as your application is not meant for old devices that support just plain HTML, or your end users are not going to rely on screen readers. If your application needs to support the C-level devices and/or users rely on screen readers, then you would want to consider where you place the panel markup, as this would be read out to the users. There is talk that this restriction could be removed in a future version of the framework, but at the time of writing this book, this constraint must be followed.

Let's add a panel to our page. We believe that the best place for the panel is after the header and before the content. This way, we can even account for users relying on screen readers and support users using the C-level devices. We will use the panel to hold our contact information. Open up the file we were just working on, if you have closed it, and add the following code after our footer `div` element:

```
<div data-role="panel" id="contactpanel" data-display="push"
    data-dismissible="true" data-theme="a">
    <div>
        <p>Contact Us!</p>
        <a href="tel:555-555-5555">(555)555-5555</a>
        <p>1234 First Avenue</p>
        <p>Anytown, Anystate 12345</p>
        <a
            href="mailto:contact@anytownciviccenter.com">
                contact@anytownciviccenter.com</a>
    </div>
</div><!-- /panel -->
```

Your file will now look like this:

```
<!DOCTYPE html>
<html>
    <head>
        <title>Anytown Civic Center</title>
        <meta name="viewport" content="width=device-width,
            minimum-scale=1.0,maximum-scale=1.0,
                user-scrolable=no">
        <link rel="stylesheet"
            href="http://code.jquery.com/mobile/1.4.5/
                jquery.mobile-1.4.5.min.css" />
        <script src="http://code.jquery.com/
            jquery-1.11.0.min.js"></script>
        <script
            src="http://code.jquery.com/mobile/1.4.5/
                jquery.mobile-1.4.5.min.js"></script>
    </head>
    <body>
        <div data-role="page">
            <div data-role="header">
                <h1>Welcome to Anytown Civic Center!</h1>
            </div><!-- /header -->

            <div data-role="panel" id="contactpanel"
                data-display="push" data-dismissible="true"
                    data-theme="a">
                <div>
                    <p>Contact Us!</p>
                    <a href="tel:555-555-5555">(555)555-5555</a>
                    <p>1234 First Avenue</p>
                    <p>Anytown, Anystate 12345</p>
                    <a href=
                        "mailto:contact@anytownciviccenter.com">
                            contact@anytownciviccenter.com</a>
                </div>
            </div><!-- /panel -->

            <div class="ui-content" role="main">
                <p>Welcome to the website for Anytown Civic
                    Center. We have a lot of great upcoming events
                        for you to check out.</p>
            </div><!-- /content -->
```

```
            <div data-role="footer">
                <h2>
                    <i>1234 First Avenue Anytown, Anystate 12345
                        USA</i>
                </h2>
            </div>
        </div><!-- /page -->
    </body>
</html>
```

Let's look at the code we added:

```
<div data-role="panel" id="contactpanel" data-display="push"
    data-dismissible="true" data-theme="a">
```

We are declaring our panel and giving it the ID of `contactpanel`. We are telling the framework that we want this panel to be a push panel and be styled by the a swatch of our theme, which happens to be the default theme that ships with jQuery Mobile.

Now, if you run this code, the page will look the same. "Why is that?" you ask. Well, we have to provide a means to make the panel visible. It remains offscreen until we execute the code to make it visible. You can do a couple of things to make it visible. You can use a link with the `href` set to the `id` element of `div` of the panel, or you can use a button to make it visible. The button looks better in our opinion, so this will be the method we will use. Change the footer section to look like the following code:

```
            <div data-role="footer">
                <h2>
                    <a href="#contactpanel" class="ui-btn"
                        data-rel="panel">Contact us</a>
                </h2>
            </div>
```

What we are doing here is creating a jQuery Mobile button with the link pointing to our panel `div` element. Now, execute the code and push the button, and you should see the following screenshot:

Contact Us!

(555)555-5555

1234 First Avenue

Anytown, Anystate 12345

contact@anytownciviccenter.com

Looking good so far. To close the panel, you can add another button to the panel `div` element that closes it, or you can simply click on the page itself to go back to the main website. We'll be adding more to this panel as we progress through the project; for now, let's move on to popups.

Popups

Popups have been around for quite some time in one aspect or another. jQuery Mobile provides many implementations of popups that we can make use of. These different types of popups include:

- Basic pop up
- Image lightbox
- Tooltip
- Modal dialog

In addition to these types of popups, the basic popup can be modified to hold menus, maps, and more. jQuery Mobile does an excellent job of providing a very powerful and robust popup system.

Over the course of this book, we will make use of a couple of the different types of popups, but for now, let's make a modal dialog letting the user know that we don't have any current events scheduled for them to register for.

Declaring one event is just as straightforward as the other jQuery Mobile widgets we've already looked at. To create a popup, add the data-role of popup to a `div` element that holds the pop-up contents. A link needs to be created with the `href` set to the `id` element of the popup, `div`, and add the `data-rel` attribute with the value `popup`. A popup, `div`, has to be nested inside the same page as that of the link.

Open up `index.html` and add the following code after our page declaration and before our header declaration:

```
<div data-role="popup" id="eventregister"
    data-overlay-theme="b" data-theme="b"
        data-dismissible="false">
    <div data-role="header" data-theme="a">
        <h1>Current Events</h1>
    </div>
```

```
<div role="main" class="ui-content">
    <h3 class="ui-title">Sorry, currently there are no
        scheduled events.</h3>
    <a href="#" class="ui-btn ui-corner-all ui-shadow
        ui-btn-inline ui-btn-b" data-rel="back">Close</a>
</div>
</div>
```

Now, we need to code in some way to make our popup appear. We'll use another button to do this; add the following code in our current content area after `<p>Welcome to the website for …`:

```
<a href="#eventregister" data-rel="popup"
    data-position-to="window" data-transition="pop" class="ui-btn
        ui-corner-all ui-shadow ui-btn-inline ui-btn-b">View
            Current Events</a>
```

Now, if you launch the website, you will first see this:

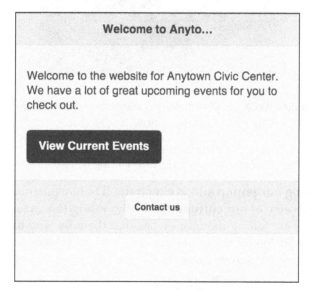

Looking good so far. Now, click on the **View Current Events** button and you will see this:

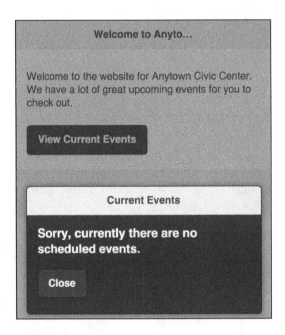

Awesome! We just created a very simple, yet elegant modal dialog box with jQuery Mobile. Let's take a look at the code to see how it was done:

```
<div data-role="popup" id="eventregister" data-overlay-theme="b"
    data-theme="b" data-dismissible="false">
```

Here, we are creating our popup and giving it the ID of eventregister. We are assigning it the b swatch of our current theme. By setting the data-dismissible option to false, we are telling the framework that the only way to get rid of the popup is to use the button on the modal itself. If we had set that to true, simply touching anywhere on the screen or using the *Esc* key would dismiss it:

```
<div data-role="header" data-theme="a">
    <h1>Current Events</h1>
</div>
```

Here, we are creating the header part of our popup. You may be saying right now, "Hey, we already have a header" and you would be right. That's the good thing about the jQuery framework; we can set and use different headers for our widgets. If we didn't have the header set here, Current Events would not be stylized in the popup and it wouldn't look right. Go ahead and remove the header declaration (the data-role="header" piece) for now, and this is what you will get:

Looks horrible, doesn't it? Put the header code back and let's continue looking at the code:

```
<div role="main" class="ui-content">
    <h3 class="ui-title">Sorry, currently there are no
        scheduled events.</h3>
    <a href="#" class="ui-btn ui-corner-all ui-shadow
        ui-btn-inline ui-btn-b" data-rel="back">Close</a>
</div>
```

This code is the content of our popup. We are applying some of jQuery Mobile's built-in CSS styling to a few of the elements, and then lastly, we will create a button that will be used to close the popup. Remember that this is the only way we can close the popup, since we set the `data-dismissible` attribute to `false`.

That's it for the popup itself, so now let's look at the button we will use to open it:

```
<a href="#eventregister" data-rel="popup"
    data-position-to="window" data-transition="pop" class="ui-btn
        ui-corner-all ui-shadow ui-btn-inline ui-btn-b">
    View Current Events
</a>
```

Here, we are creating a jQuery Mobile button and passing the `href` element the ID of the `div` element it needs to open. You can have several popups on a page, so please be sure to keep them straight with your `div` IDs. For `data-transition`, we are telling it to pop onto the screen, which will have it appear extremely fast with little fanfare.

Data transitions

There are 10 different data transitions that you can use on popups. These transitions are as follows:

- No transition
- Pop
- Fade
- Flip
- Turn
- Flow
- Slide
- Slidefade
- Slide up
- Slide down

Each of them has a different effect when bringing the popup; play around with the one we just created and try out these different transitions, so you can see each of them in action and find the one that you may like.

Here is what our overall code should now look like so far; in this way, if something isn't working for you anymore, you can refer to the code to make sure everything has been done correctly:

```
<!DOCTYPE html>
<html>
    <head>
        <title>Anytown Civic Center</title>
        <meta name="viewport" content="width=device-width,
            minimum-scale=1.0,maximum-scale=1.0,
                user-scrolable=no">
        <link rel="stylesheet"
            href="http://code.jquery.com/mobile/1.4.5/
                jquery.mobile-1.4.5.min.css" />
        <script src="http://code.jquery.com/
            jquery-1.11.0.min.js"></script>
        <script
            src="http://code.jquery.com/mobile/1.4.5/
                jquery.mobile-1.4.5.min.js"></script>
    </head>
    <body>
        <div data-role="page">
            <div data-role="popup" id="eventregister"
                data-overlay-theme="b" data-theme="b"
                    data-dismissible="false">
```

```
<div data-role="header" data-theme="a">
    <h1>Current Events</h1>
</div>
<div role="main" class="ui-content">
    <h3 class="ui-title">Sorry, currently there
        are no scheduled events.</h3>
    <a href="#" class="ui-btn ui-corner-all
        ui-shadow ui-btn-inline ui-btn-b"
            data-rel="back">Close</a>
</div>
</div>

<div data-role="header">
    <h1>Welcome to Anytown Civic Center!</h1>
</div><!-- /header -->

<div data-role="panel" id="contactpanel"
    data-display="push" data-dismissible="true"
        data-theme="a">
    <div>
        <p>Contact Us!</p>
        <a href="tel:555-555-5555">(555)555-5555</a>
        <p>1234 First Avenue</p>
        <p>Anytown, Anystate 12345</p>
        <a href=
            "mailto:contact@anytownciviccenter.com">
                contact@anytownciviccenter.com</a>
    </div>
</div><!-- /panel -->

<div class="ui-content" role="main">
    <p>Welcome to the website for Anytown Civic
        Center. We have a lot of great upcoming events
            for you to check out.</p>
    <a href="#eventregister" data-rel="popup"
        data-position-to="window"
            data-transition="pop" class="ui-btn
                ui-corner-all ui-shadow ui-btn-inline
                    ui-btn-b">
        View Current Events
    </a>
</div><!-- /content -->

<div data-role="footer">
    <h2>
        <a href="#contactpanel" class="ui-btn
            ui-corner-all" data-rel="panel">Contact
                us</a>
```

```
                  </h2>
              </div>
         </div><!-- /page -->
      </body>
  </html>
```

Now that we are done with popups and everything appears to be working, let's move on to our next section.

Toolbars

Toolbars in jQuery Mobile are used to enhance the headers and footers of the pages. We have already used headers and footers in our page, so now it is time to make them a bit fancier and add some functionality to our app.

Toolbars is probably one of the easiest jQuery Mobile widgets to implement, especially if you already have a header or footer. To show how easy this is, we are going to change our existing header and footer sections into toolbars.

Before we get started, let's jot down our plan:

1. Create a Home button in our header.
2. Move our Events button from the content to our header.
3. Stylize our Contact button in our footer to look like a toolbar button.

This doesn't seem like a whole lot of stuff to do, but it will make a beautiful difference in our mobile app. To begin, we'll create a Home button.

Creating a Home button

Keeping with the trend, we've been going with on this chapter, open up index.html and add the following code right under our initial header declaration:

```
<a href="index.html" class="ui-btn-left ui-btn ui-btn-inline
    ui-mini ui-corner-all ui-btn-b ui-btn-icon-left
       ui-icon-home">Home</a>
```

Save the code and view it, you should see this:

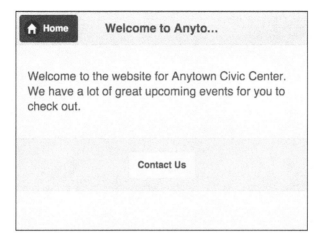

Simple, right? Let's talk about what this code does:

```
<a href="index.html" class="ui-btn-left ui-btn ui-btn-inline
    ui-mini ui-corner-all ui-btn-b ui-btn-icon-left
        ui-icon-home">Home</a>
```

We are creating a link to our main page, `index.html` (for now) by adding a button that has an icon in it. We have already seen how we style the link tag like a button using the jQuery Mobile specific button classes in the example so far. A few new classes here include:

- `ui-btn-left`: This class informs the framework that the button is to be aligned to the left end of the toolbar
- `ui-mini`: This informs the framework to render a smaller size button
- `ui-btn-b`: This informs the framework to render a swatch b styled button
- `ui-btn-icon-left`: This informs the framework that the icon that is to be used will be aligned to the left-hand side of the button
- `ui-icon-home`: This informs the framework to render a **Home** icon

Now, we are going to move the current **View Current Events** button. First thing we need to do is go ahead and delete the current button. Now, add the following code right before the closing of our header `div`:

```
<a href="#eventregister" data-rel="popup"
    data-position-to="window" data-transition="pop"
        class="ui-btn-right ui-btn-b ui-btn ui-btn-inline ui-mini
            ui-corner-all ui-btn-icon-right
                ui-icon-calendar">Events</a>
```

Save the code and you will now see this:

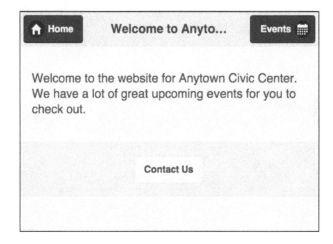

Our mobile app is coming along nicely, isn't it? Now, it's time to analyze the code again:

```
<a href="#eventregister" data-rel="popup"
    data-position-to="window" data-transition="pop"
        class="ui-btn-right ui-btn-b ui-btn ui-btn-inline ui-mini
            ui-corner-all ui-btn-icon-right
                ui-icon-calendar">Events</a>
```

Just like our previous button, we are creating a link to our popup and using the calendar icon for our button. The rest of the code is similar to what we used in our original popup declaration and the icon button we just used for the Home link. This brings out the true power of this awesome framework, wherein we can combine different attributes and features of this framework into one single unit to enhance our mobile web application.

We now have a fairly nice header toolbar. In the coming chapters, we will end up changing this header, but it does the job for now. Now, let's create our footer toolbar.

Remove the current contact button we have in our footer and replace it with this:

```
<a href="#contactpanel" class="ui-btn ui-btn-inline ui-mini
    ui-corner-all ui-btn-b ui-btn-icon-left ui-icon-phone">Contact
        Us</a>
```

Save this and when you view the project, you should see something like this:

It's quite amazing what a few lines of code can do to a mobile application in jQuery Mobile. Let's finish this section off by looking at that last bit of code we added:

```
<a href="#contactpanel" data-rel="panel" class="ui-btn
    ui-btn-inline ui-mini ui-corner-all ui-btn-b ui-btn-icon-left
        ui-icon-phone">Contact Us</a>
```

We are doing the same thing what we did with the other two buttons. We're creating a link to our panel and assigning the phone icon to it.

That is it for toolbars and here is what our code currently looks like:

```
<!DOCTYPE html>
<html>
    <head>
        <title>Anytown Civic Center</title>
        <meta name="viewport" content="width=device-width,
            minimum-scale=1.0,maximum-scale=1.0,
                user-scrolable=no">
        <link rel="stylesheet"
            href="http://code.jquery.com/mobile/1.4.5/
                jquery.mobile-1.4.5.min.css" />
        <script src="http://code.jquery.com/jquery-1.11.0.min.js">
            </script>
        <script
            src="http://code.jquery.com/mobile/1.4.5/
                jquery.mobile-1.4.5.min.js"></script>
    </head>
    <body>
        <div data-role="page">
```

```
<div data-role="popup" id="eventregister"
    data-overlay-theme="b" data-theme="b"
        data-dismissible="false">
    <div data-role="header" data-theme="a">
        <h1>Current Events</h1>
    </div>
    <div role="main" class="ui-content">
        <h3 class="ui-title">Sorry, currently there
            are no scheduled events.</h3>
        <a href="#" class="ui-btn ui-corner-all
            ui-shadow ui-btn-inline ui-btn-b"
                data-rel="back">Close</a>
    </div>
</div>


<div data-role="header">
    <a href="index.html" class="ui-btn-left ui-btn
        ui-btn-inline ui-mini ui-corner-all ui-btn-b
            ui-btn-icon-left ui-icon-home">Home</a>
    <h1>Welcome to Anytown Civic Center!</h1>
    <a href="#eventregister" data-rel="popup"
        data-position-to="window"
            data-transition="pop" class="ui-btn-right
                ui-btn-b ui-btn ui-btn-inline ui-mini
                    ui-corner-all ui-btn-icon-right
                        ui-icon-calendar">Events</a>
</div><!-- /header -->


<div data-role="panel" id="contactpanel"
    data-display="push" data-dismissible="true"
        data-theme="a">
    <div>
        <p>Contact Us!</p>
        <a href="tel:555-555-5555">(555)555-5555</a>
        <p>1234 First Avenue</p>
        <p>Anytown, Anystate 12345</p>
        <a href=
            "mailto:contact@anytownciviccenter.com">
                contact@anytownciviccenter.com</a>
    </div>
</div><!-- /panel -->


<div class="ui-content" role="main">
    <p>Welcome to the website for Anytown Civic
        Center. We have a lot of great upcoming events
            for you to check out.</p>
```

```
        </div><!-- /content -->

        <div data-role="footer">
            <h2>
                <a href="#contactpanel" data-rel="panel"
                    class="ui-btn ui-btn-inline ui-mini
                        ui-corner-all ui-btn-b
                            ui-btn-icon-left
                                ui-icon-phone">Contact Us</a>
            </h2>
        </div>
    </div><!-- /page -->
</body>
</html>
```

 We used several of the built-in icons for jQuery Mobile in this chapter. There is plenty more. You can find the full list at http://demos.jquerymobile.com/1.4.5/icons/.

Navbars

Navbars is another great widget that is featured in jQuery Mobile. It is a simple navbar but can also be turned into a tab bar that you see in the native mobile applications. It supports up to five buttons and those buttons can also use the icons we used in the last section.

In simple terms, a jQuery Mobile navbar is an unordered list that is wrapped within a div container with data-role of navbar. You can create them in the header or footer of your page.

We are going to add one to our header div element that will serve as our navigation menu in the later chapters. This means for now that it won't be very functional, but the groundwork will be there and we can still see the navbar in action to an extent.

Open up index.html and add the following code before our closing header div element:

```
<div data-role="navbar">
    <ul>
        <li><a href="#">About Us</a></li>
        <li><a href="#">Facilities</a></li>
        <li><a href="#">Catering</a></li>
    </ul>
</div><!-- /navbar -->
```

Save the file and run the code. You should see the following screenshot:

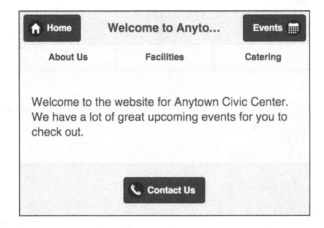

Go ahead and click on the links. The active button will turn blue as you click on it. And now, it's code review time!

```
<div data-role="navbar">
        <ul>
            <li><a href="#">About Us</a></li>
            <li><a href="#">Facilities</a></li>
            <li><a href="#">Catering</a></li>
        </ul>
    </div><!-- /navbar -->
```

We told you that, all that a navbar is; it's an unordered list wrapped within the navbar `data-role` container. As you add more elements (remember the limit is five), jQuery Mobile automatically resizes the bar so that the buttons fit comfortably on the screen. If we were to put this code in the footer `div` element, it would act almost like a native mobile application tab bar.

As previously mentioned, you can use icons in a navbar via the `data-icon` attribute on the links that we used in the toolbar. Go ahead and practice with that, as well as seeing what this looks like in the footer.

This concludes our code for this chapter. For one last review, in case anything isn't working correctly in your code, here is the full and final code for this chapter. It is important that your code is functioning properly, since this code will form the basis of the rest of the book:

```
<!DOCTYPE html>
<html>
    <head>
```

```
    <title>Anytown Civic Center</title>
    <meta name="viewport" content="width=device-width,
        minimum-scale=1.0,maximum-scale=1.0,
            user-scrolable=no">
    <link rel="stylesheet"
        href="http://code.jquery.com/mobile/1.4.5/
            jquery.mobile-1.4.5.min.css" />
    <script src="http://code.jquery.com/
        jquery-1.11.0.min.js"></script>
    <script
        src="http://code.jquery.com/mobile/1.4.5/
            jquery.mobile-1.4.5.min.js"></script>
</head>
<body>
    <div data-role="page">
        <div data-role="popup" id="eventregister"
            data-overlay-theme="b" data-theme="b"
                data-dismissible="false">
            <div data-role="header" data-theme="a">
                <h1>Current Events</h1>
            </div>
            <div role="main" class="ui-content">
                <h3 class="ui-title">Sorry, currently there
                    are no scheduled events.</h3>
                <a href="#" class="ui-btn ui-corner-all
                    ui-shadow ui-btn-inline ui-btn-b"
                        data-rel="back">Close</a>
            </div>
        </div>

        <div data-role="header">
            <a href="index.html" class="ui-btn-left ui-btn
                ui-btn-inline ui-mini ui-corner-all ui-btn-b
                    ui-btn-icon-left ui-icon-home">Home</a>
            <h1>Welcome to Anytown Civic Center!</h1>
            <a href="#eventregister" data-rel="popup"
                data-position-to="window"
                    data-transition="pop" class="ui-btn-right
                        ui-btn-b ui-btn ui-btn-inline ui-mini
                            ui-corner-all ui-btn-icon-right
                                ui-icon-calendar">Events</a>
            <div data-role="navbar">
                <ul>
                    <li><a href="#">About Us</a></li>
                    <li><a href="#">Facilities</a></li>
                    <li><a href="#">Catering</a></li>
```

```
                </ul>
            </div><!-- /navbar -->
        </div><!-- /header -->

        <div data-role="panel" id="contactpanel"
            data-display="push" data-dismissible="true"
                data-theme="a">
            <div>
                <p>Contact Us!</p>
                <a href="tel:555-555-5555">(555)555-5555</a>
                <p>1234 First Avenue</p>
                <p>Anytown, Anystate 12345</p>
                <a href=
                    "mailto:contact@anytownciviccenter.com">
                        contact@anytownciviccenter.com</a>
            </div>
        </div><!-- /panel -->

        <div class="ui-content" role="main">
            <p>Welcome to the website for Anytown Civic
                Center. We have a lot of great upcoming events
                    for you to check out.</p>
        </div><!-- /content -->

        <div data-role="footer">
            <h2>
                <a href="#contactpanel" data-rel="panel"
                    class="ui-btn ui-btn-inline ui-mini
                        ui-corner-all ui-btn-b
                            ui-btn-icon-left
                                ui-icon-phone">Contact Us</a>
            </h2>
        </div>
    </div><!-- /page -->
    </body>
</html>
```

Summary

This chapter was very code intensive, but at the same time, the code was simple, yet elegant and powerful. This is one of the great things about jQuery Mobile. It allows us to make beautiful and functional mobile web applications with just a few lines of code.

Let's review the topics we covered in this chapter. This chapter focused on some of the most popular and important widgets of jQuery Mobile. We saw that the page widget is the main widget required for the rest of the widgets. We saw how panels can be used to display information and how the popup can be powerful enough to hold an image or form, or serve as a full-blown modal dialog. We rounded off the chapter by seeing how a toolbar can add some nice functionality to our app, and then how the navbar will serve as a nice menu system for later chapters.

Remember to check your code to make sure it is working. We know that we have said this a great deal, but this chapter provides the framework for the rest of the book. If your code isn't functioning properly now, you will have some difficulty in the later parts of the book.

In the next chapter, we will start putting together a navigation system for our application as we link the pages together. We will also explore the touch interactions available in the framework to swipe through the pages we created in this chapter.

5
Navigation

In the previous chapter, we looked at creating pages with jQuery Mobile. However, we didn't really look at ways in which you could navigate through these pages. In this chapter, we will look at several different navigation methods that are included as part of the framework.

Overview

The ease of navigation on a mobile device can really make or break your website or application. Luckily, jQuery Mobile includes some techniques that can not only make the navigation easy, but make it look spectacular as well. In this chapter, we will be looking at the following topics:

- Navigating between pages
- Transitions and effects
- Navigating with swipe gestures
- Working with page loaders

Of course, we will continue working on our civic center example project, so make sure it is loaded up. Make sure that you have the code mentioned at the end of the previous chapter.

Links and page navigation

When the jQuery Mobile framework encounters a link, it loads up the link in one of the following two ways: Ajax or non-Ajax.

Ajax page linking

By default, whenever you create a link, the framework will automatically load the link using Ajax. By using Ajax, this allows you to have animated page transitions. Again, this is by default, so you don't have to do anything to your link. The framework parses the anchor tag's `href` attribute and automatically loads the target into the DOM. If it fails, an error message stating **Error Loading Page** will display briefly on the screen.

Let's go ahead and see an Ajax link in action:

1. Go to your project and copy (don't move or rename) `index.html` to a new file named `about.html`.

2. Next, edit the link we have for **About Us** to `about.html` (remember to do this in the `index.html` file as well).

3. Now, change the `content` div element (`<div data-role="content">`) of `about.html` to say this:

   ```
   Founded in 2014, Anytown Civic Center plans to be the
       premier venue for the hottest concerts, the greatest
           sports events, the best conferences and the most
               extravagant weddings.
   ```

4. Save the file and load up the project in your browser of choice.

5. Click on the **About Us** link.

Did you notice the brief page flicker before it loaded the about page? This is a default page animation that is executed when the user clicks on our link.

In case you need it, here is what the body of `about.html` should look like:

```
<div data-role="page">
    <div data-role="popup" id="eventregister"
        data-overlay-theme="b" data-theme="b"
            data-dismissible="false">
        <div data-role="header" data-theme="a">
            <h1>Current Events</h1>
        </div>
        <div role="main" class="ui-content">
```

```
        <h3 class="ui-title">Sorry, currently there are no
            scheduled events.</h3>
        <a href="#" class="ui-btn ui-corner-all ui-shadow
            ui-btn-inline ui-btn-b" data-rel="back">Close</a>
    </div>
</div>

<div data-role="header">
    <a href="index.html" class="ui-btn-left ui-btn
        ui-btn-inline ui-mini ui-corner-all
            ui-btn-b ui-btn-icon-left ui-icon-home">Home</a>
    <h1>Welcome to Anytown Civic Center!</h1>
    <a href="#eventregister" data-rel="popup"
        data-position-to="window" data-transition="pop"
            class="ui-btn-right ui-btn-b ui-btn ui-btn-inline
                ui-mini ui-corner-all ui-btn-icon-right
                    ui-icon-calendar">Events</a>
    <div data-role="navbar">
        <ul>
            <li><a href="about.html">About Us</a></li>
            <li><a href="#">Facilities</a></li>
            <li><a href="#">Catering</a></li>
        </ul>
    </div><!-- /navbar -->
</div><!-- /header -->

<div data-role="panel" id="contactpanel" data-display="push"
    data-dismissible="true" data-theme="a">
    <div>
        <p>Contact Us!</p>
        <a href="tel:555-555-5555">(555)555-5555</a>
        <p>1234 First Avenue</p>
        <p>Anytown, Anystate 12345</p>
        <a href="mailto:contact@anytownciviccenter.com">
            contact@anytownciviccenter.com</a>
    </div>
</div><!-- /panel -->

<div class="ui-content" role="main">
    <p>Founded in 2014, Anytown Civic Center plans to be the
        premier venue for the hottest concerts, the greatest
            sports events, the best conferences and the most
                extravagant weddings.</p>
</div><!-- /content -->
```

```
<div data-role="footer">
    <h2>
        <a href="#contactpanel" data-rel="panel" class="ui-btn
            ui-btn-inline ui-mini ui-corner-all ui-btn-b
                ui-btn-icon-left ui-icon-phone">Contact Us</a>
    </h2>
</div>
</div><!-- /page -->
```

We are not going to analyze the code this time, as we did nothing but just changed some text on the screen.

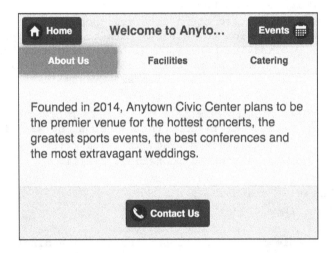

Non-Ajax page linking

For security reasons, by default, every external link (links that leave the current domain) are executed without Ajax, but you can also specify specific links to be loaded without Ajax by including one of the following attributes in your href declaration:

- rel="external"
- data-ajax="false"
- target

Whenever the framework encounters a link with any of the preceding code, the link will be loaded without Ajax; therefore, there will be no animation or transition as you would encounter when using Ajax and the page will perform a full refresh.

Let's try this out so that we can compare the results of Ajax versus non-Ajax.

1. Create a new file named `facilities.html` and copy the contents of `index.html` over to it, just like we did in the previous section.

2. Change the link for `Facilities` to `facilities.html` in each file and add the `data-ajax="false"` code to the link, so that it looks like this:

```
<li><a href="facilities.html"
    data-ajax="false">Facilities</a></li>
```

3. Make the changes in the following code so that your `content` div element looks like this:

```
<div class="ui-content" role="main">
    <p>At Anytown Civic Center we have the following
        facilities at your disposal:<br />
        <ul>
            <li>3 Banquet Halls</li>
            <li>1 Sports Arena</li>
            <li>25 Conference Rooms of various
                sizes</li>
            <li>2 Ballrooms</li>
        </ul>
        Contact us for pricing and availability.
    </p>
</div><!-- /content -->
```

4. Save the file and load up the project in your browser of choice, and click on the **Facilities** link.

Did you notice the difference? There was no spiffy little transition this time. There was just a page refresh, and then our page. You may be asking yourself, "Why would we ever use this on an internal page?" Well, there is a good reason, which we'll discuss in the next section. The `facilities.html` page will look as seen in the following screenshot:

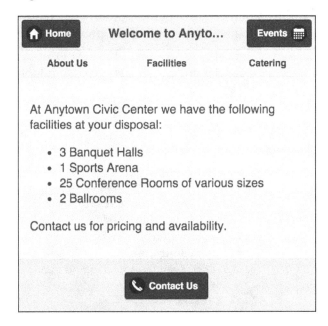

Multipage document linking

Technically, there is a third methodology for page navigation within jQuery Mobile. jQuery Mobile supports multiple data-roles of the type `page` within one file. This means you could build your entire web application in one file; the framework simply loads the first page role it comes to as the first page in the application.

To link to other "pages" within the file, we link to the page via an anchor that corresponds to the ID of the page role we want to link to. This sounds extremely useful; we could have every page in one file and just have that one file to maintain. There is one nuance to this method though; the framework does not use Ajax with these links if you have a mix of multipage and single page files. That means you will not have any animations or page loaders. If you build your entire application using the single page implementation technique, then you will be okay.

Let's take a look at this in action:

We will add a new data-role of the type `page` in the `index.html` file, right after where the `div` element for the existing `div` container with `data-role="page"` ends. For ease of coding, copy the entire `div` element of `data-role="page"` for our index page and paste it right after where it ends. So, now you will basically have two exactly same `div` elements with data-role of the type `page`.

We will see in detail what our catering `div` element would look like in just a couple of minutes, but before that, we need to make a small change in the current `page` div element:

1. Add `id="home"` to the existing `div` element with data-role of the type `page`. The `div` element will now look like the following code:

    ```
    <div data-role="page" id="home">
    ```

    We will now be modifying the newly added page to represent our `catering` details.

2. Make the changes in the following code so that the content of the newly added page looks like this:

    ```
    <div class="ui-content" role="main">
        <p>
            We offer a full catering service for your events.
                Our three star Michelin Chef and his staff can
                    tackle almost any cuisine you may need for
                        your guests.<br />Contact us for a full
                            menu and prices.
        </p>
    </div><!-- /content -->
    ```

3. Now, add `id="catering"` to this newly added `div` element with `data-role="page"`. Your `div` element will now look like this:

    ```
    <div data-role="page" id="catering">
    ```

4. Add `id="eventregister2"` to the `div` element that follows with `data-role="popup"`. This `div` element will now reflect the following code:

    ```
    <div data-role="popup" id="eventregister2"
        data-overlay-theme="b" data-theme="b"
            data-dismissible="false">
    ```

5. A few lines later, in the div element with data-role="header", we will have to change the href attribute value of the Events button because of the change in ID of the popup in the preceding step. This change will result in the following code:

```
<a href="#eventregister2" data-rel="popup"
    data-position-to="window" data-transition="pop"
        class="ui-btn-right ui-btn-b ui-btn ui-btn-inline
            ui-mini ui-corner-all ui-btn-icon-right
                ui-icon-calendar">Events</a>
```

6. In the same header div, just a couple of lines above, update the href attribute of the Home button link to reflect a relative path to the home page with id="home". Make sure you make this change in the original page with id="home":

```
<a href="index.html#home" class="ui-btn-left ui-btn
    ui-btn-inline ui-mini ui-corner-all ui-btn-b
        ui-btn-icon-left ui-icon-home">Home</a>
```

7. Now, update the link to the catering page to reference our catering div element. If all of this was in one file, we could simply put the anchor ID here instead of the absolute link, but since we are mixing internal, external, and multipage, it is in our best interest to do it this way. Make sure that you perform this change on the original page with id="home". This change in the code will be reflected in the following line:

```
<li><a href="index.html#catering">Catering</a></li>
```

Though simple, that's a lot of changes we have made in the code. Just to make sure you have not missed any of these, here is what your catering div element should look like:

```
<div data-role="page" id="catering">
    <div data-role="popup" id="eventregister2"
        data-overlay-theme="b" data-theme="b"
            data-dismissible="false">
        <div data-role="header" data-theme="a">
            <h1>Current Events</h1>
        </div>
        <div role="main" class="ui-content">
            <h3 class="ui-title">Sorry, currently there are no
                scheduled events.</h3>
            <a href="#" class="ui-btn ui-corner-all ui-shadow
                ui-btn-inline ui-btn-b" data-rel="back">Close</a>
        </div>
    </div>
```

```
<div data-role="header">
    <a href="index.html#home" class="ui-btn-left ui-btn
        ui-btn-inline ui-mini ui-corner-all ui-btn-b
            ui-btn-icon-left ui-icon-home">Home</a>
    <h1>Welcome to Anytown Civic Center!</h1>
    <a href="#eventregister2" data-rel="popup"
        data-position-to="window" data-transition="pop"
            class="ui-btn-right ui-btn-b ui-btn ui-btn-inline
                ui-mini ui-corner-all ui-btn-icon-right
                    ui-icon-calendar">Events</a>
    <div data-role="navbar">
        <ul>
            <li><a href="about.html">About Us</a></li>
            <li><a href="facilities.html" data-ajax="false">
                Facilities</a></li>
            <li><a href="index.html#catering">Catering</a>
                </li>
        </ul>
    </div><!-- /navbar -->
</div><!-- /header -->

<div data-role="panel" id="contactpanel" data-display="push"
    data-dismissible="true" data-theme="a">
    <div>
        <p>Contact Us!</p>
        <a href="tel:555-555-5555">(555)555-5555</a>
        <p>1234 First Avenue</p>
        <p>Anytown, Anystate 12345</p>
        <a href="mailto:contact@anytownciviccenter.com">
            contact@anytownciviccenter.com</a>
    </div>
</div><!-- /panel -->

<div class="ui-content" role="main">
    <p>
        We offer a full catering service for your events. Our
            three star Michelin Chef and his staff can tackle
                almost any cuisine you may need for your
                    guests.<br />Contact us for a full menu
                        and prices.
    </p>
</div><!-- /content -->
```

```
    <div data-role="footer">
        <h2>
            <a href="#contactpanel" data-rel="panel" class="ui-btn
                ui-btn-inline ui-mini ui-corner-all ui-btn-b
                    ui-btn-icon-left ui-icon-phone">Contact Us</a>
        </h2>
    </div>
</div><!-- /page -->
```

Right off the bat, you can tell this file has doubled in length. This is because we are creating another page within this file and it has the same header/footer as the other page. We have our catering page ready. It is just like our other pages, with some changes for links and content. Save this file and be sure to update your links to these pages in your other files. Once you do that, load it up and try it out. Our catering page will look as shown in the following screenshot:

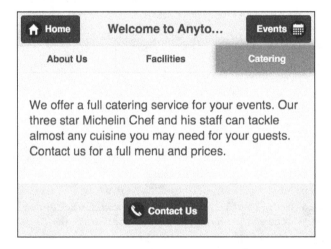

"Hey, wait a minute guys, I see transitions and a page loader. You said multipage linking wouldn't have that!" You are correct, we did say that, but we are not doing true multipage linking here. Our links, such as index.html#home, get treated as normal links. This allows us to navigate through multipage, internal, and external pages all seamlessly.

The only caveat you should be aware of is that when linking to a multipage document, you need to link them as we did on the non-Ajax links when going from a single page to a multipage; in this way, it will do a full page refresh so that it will function properly.

After you have tested this out, let's go ahead and put all the pages into a single file; in this way, we can have one file to maintain and still use animations and page loaders. So, copy the page data-roles out of each of our files and put them all in `index.html` in this order (with these IDs):

- `home`
- `about`
- `facilities`
- `catering`

Update all your links well to link to the correct anchor. Your `index.html` body should reflect the `div` elements with the data-role page in the following sequence.

First of all comes the `Home` page:

```
<!-- Home page begins -->
    <div data-role="page" id="home">
        <div data-role="popup" id="eventregister"
            data-overlay-theme="b" data-theme="b"
                data-dismissible="false">
            <div data-role="header" data-theme="a">
                <h1>Current Events</h1>
            </div>
            <div role="main" class="ui-content">
                <h3 class="ui-title">Sorry, currently there are no
                    scheduled events.</h3>
                <a href="#" class="ui-btn ui-corner-all ui-shadow
                    ui-btn-inline ui-btn-b"
                        data-rel="back">Close</a>
            </div>
        </div>

        <div data-role="header">
            <a href="#home" class="ui-btn-left ui-btn
                ui-btn-inline ui-mini ui-corner-all ui-btn-b
                    ui-btn-icon-left ui-icon-home">Home</a>
            <h1>Welcome to Anytown Civic Center!</h1>
            <a href="#eventregister" data-rel="popup"
                data-position-to="window" data-transition="pop"
                    class="ui-btn-right ui-btn-b ui-btn
                        ui-btn-inline ui-mini ui-corner-all
                            ui-btn-icon-right
                                ui-icon-calendar">Events</a>
            <div data-role="navbar">
                <ul>
```

```
                <li><a href="#about">About Us</a></li>
                <li><a href="#facilities"
                    data-ajax="false">Facilities</a></li>
                <li><a href="#catering">Catering</a></li>
            </ul>
        </div><!-- /navbar -->
    </div><!-- /header -->

    <div data-role="panel" id="contactpanel"
        data-display="push" data-dismissible="true"
            data-theme="a">
        <div>
            <p>Contact Us!</p>
            <a href="tel:555-555-5555">(555)555-5555</a>
            <p>1234 First Avenue</p>
            <p>Anytown, Anystate 12345</p>
            <a href="mailto:contact@anytownciviccenter.com">
                contact@anytownciviccenter.com</a>
        </div>
    </div><!-- /panel -->

    <div class="ui-content" role="main">
        <p>Welcome to the website for Anytown Civic Center. We
            have a lot of great upcoming events for you to
                check out.</p>
    </div><!-- /content -->

    <div data-role="footer">
        <h2>
            <a href="#contactpanel" data-rel="panel"
                class="ui-btn ui-btn-inline ui-mini
                    ui-corner-all ui-btn-b ui-btn-icon-left
                        ui-icon-phone">Contact Us</a>
        </h2>
    </div>
    </div><!-- /page -->
<!-- Home page ends -->
```

This will be followed by the About page:

```
<!-- About page begins -->
    <div data-role="page" id="about">
        <div data-role="popup" id="eventregister2"
            data-overlay-theme="b" data-theme="b"
                data-dismissible="false">
```

```
<div data-role="header" data-theme="a">
    <h1>Current Events</h1>
</div>
<div role="main" class="ui-content">
    <h3 class="ui-title">Sorry, currently there are no
        scheduled events.</h3>
    <a href="#" class="ui-btn ui-corner-all ui-shadow
        ui-btn-inline ui-btn-b"
            data-rel="back">Close</a>
</div>
</div>

<div data-role="header">
    <a href="#home" class="ui-btn-left ui-btn
        ui-btn-inline ui-mini ui-corner-all ui-btn-b
            ui-btn-icon-left ui-icon-home">Home</a>
    <h1>Welcome to Anytown Civic Center!</h1>
    <a href="#eventregister2" data-rel="popup"
        data-position-to="window" data-transition="pop"
            class="ui-btn-right ui-btn-b ui-btn
                ui-btn-inline ui-mini ui-corner-all
                    ui-btn-icon-right
                        ui-icon-calendar">Events</a>
    <div data-role="navbar">
        <ul>
            <li><a href="#about">About Us</a></li>
            <li><a href="#facilities"
                data-ajax="false">Facilities</a></li>
            <li><a href="#catering">Catering</a></li>
        </ul>
    </div><!-- /navbar -->
</div><!-- /header -->

<div data-role="panel" id="contactpanel"
    data-display="push" data-dismissible="true"
        data-theme="a">
    <div>
        <p>Contact Us!</p>
        <a href="tel:555-555-5555">(555)555-5555</a>
        <p>1234 First Avenue</p>
        <p>Anytown, Anystate 12345</p>
        <a href="mailto:contact@anytownciviccenter.com">
            contact@anytownciviccenter.com</a>
    </div>
</div><!-- /panel -->
```

```
      <div class="ui-content" role="main">
          <p>Founded in 2014, Anytown Civic Center plans to be
              the premier venue for the hottest concerts, the
                  greatest sports events, the best conferences
                      and the most extravagant weddings.</p>
      </div><!-- /content -->

      <div data-role="footer">
          <h2>
              <a href="#contactpanel" data-rel="panel"
                  class="ui-btn ui-btn-inline ui-mini
                      ui-corner-all ui-btn-b ui-btn-icon-left
                          ui-icon-phone">Contact Us</a>
          </h2>
      </div>
  </div><!-- /page -->
<!-- About page ends -->
```

Next to follow the About page is the Facilities page:

```
<!-- Facilities page begins-->
  <div data-role="page" id="facilities">
      <div data-role="popup" id="eventregister3"
          data-overlay-theme="b" data-theme="b"
              data-dismissible="false">
          <div data-role="header" data-theme="a">
              <h1>Current Events</h1>
          </div>
          <div role="main" class="ui-content">
              <h3 class="ui-title">Sorry, currently there are no
                  scheduled events.</h3>
              <a href="#" class="ui-btn ui-corner-all ui-shadow
                  ui-btn-inline ui-btn-b"
                      data-rel="back">Close</a>
          </div>
      </div>

      <div data-role="header">
          <a href="#home" class="ui-btn-left ui-btn
              ui-btn-inline ui-mini ui-corner-all ui-btn-b
                  ui-btn-icon-left ui-icon-home">Home</a>
          <h1>Welcome to Anytown Civic Center!</h1>
```

```
      <a href="#eventregister3" data-rel="popup"
          data-position-to="window" data-transition="pop"
              class="ui-btn-right ui-btn-b ui-btn
                  ui-btn-inline ui-mini ui-corner-all
                      ui-btn-icon-right
                          ui-icon-calendar">Events</a>
      <div data-role="navbar">
          <ul>
              <li><a href="#about">About Us</a></li>
              <li><a href="#facilities"
                  data-ajax="false">Facilities</a></li>
              <li><a href="#catering">Catering</a></li>
          </ul>
      </div><!-- /navbar -->
</div><!-- /header -->

<div data-role="panel" id="contactpanel"
    data-display="push" data-dismissible="true"
        data-theme="a">
    <div>
        <p>Contact Us!</p>
        <a href="tel:555-555-5555">(555)555-5555</a>
        <p>1234 First Avenue</p>
        <p>Anytown, Anystate 12345</p>
        <a href="mailto:contact@anytownciviccenter.com">
            contact@anytownciviccenter.com</a>
    </div>
</div><!-- /panel -->

<div class="ui-content" role="main">
    <p>At Anytown Civic Center we have the following
        facilities at your disposal:<br />
        <ul>
            <li>3 Banquet Halls</li>
            <li>1 Sports Arena</li>
            <li>25 Conference Rooms of various sizes</li>
            <li>2 Ballrooms</li>
        </ul>
        Contact us for pricing and availability.
    </p>
</div><!-- /content -->
```

```
        <div data-role="footer">
            <h2>
                <a href="#contactpanel" data-rel="panel" class="ui-btn
ui-btn-inline ui-mini ui-corner-all ui-btn-b ui-btn-icon-left ui-icon-
phone">Contact Us</a>
            </h2>
        </div>
    </div><!-- /page -->
<!-- Facilities page ends -->
```

The last is the `Catering` page, which follows the `Facilities` page:

```
<!-- Catering page begins -->
    <div data-role="page" id="catering">
        <div data-role="popup" id="eventregister4"
            data-overlay-theme="b" data-theme="b"
                data-dismissible="false">
            <div data-role="header" data-theme="a">
                <h1>Current Events</h1>
            </div>
            <div role="main" class="ui-content">
                <h3 class="ui-title">Sorry, currently there are no
                    scheduled events.</h3>
                <a href="#" class="ui-btn ui-corner-all ui-shadow
                    ui-btn-inline ui-btn-b"
                        data-rel="back">Close</a>
            </div>
        </div>

        <div data-role="header">
            <a href="#home" class="ui-btn-left ui-btn
                ui-btn-inline ui-mini ui-corner-all ui-btn-b
                    ui-btn-icon-left ui-icon-home">Home</a>
            <h1>Welcome to Anytown Civic Center!</h1>
            <a href="#eventregister4" data-rel="popup"
                data-position-to="window" data-transition="pop"
                    class="ui-btn-right ui-btn-b ui-btn
                        ui-btn-inline ui-mini ui-corner-all
                            ui-btn-icon-right
                                ui-icon-calendar">Events</a>
            <div data-role="navbar">
                <ul>
```

```
            <li><a href="#about">About Us</a></li>
            <li><a href="#facilities"
                data-ajax="false">Facilities</a></li>
            <li><a href="#catering">Catering</a></li>
        </ul>
    </div><!-- /navbar -->
</div><!-- /header -->

<div data-role="panel" id="contactpanel"
    data-display="push" data-dismissible="true"
        data-theme="a">
    <div>
        <p>Contact Us!</p>
        <a href="tel:555-555-5555">(555)555-5555</a>
        <p>1234 First Avenue</p>
        <p>Anytown, Anystate 12345</p>
        <a href="mailto:contact@anytownciviccenter.com">
            contact@anytownciviccenter.com</a>
    </div>
</div><!-- /panel -->

<div class="ui-content" role="main">
    <p>
        We offer a full catering service for your events.
            Our three star Michelin Chef and his staff can
                tackle almost any cuisine you may need for
                    your guests.<br />Contact us for a
                        full menu and prices.
    </p>
</div><!-- /content -->

<div data-role="footer">
    <h2>
        <a href="#contactpanel" data-rel="panel"
            class="ui-btn ui-btn-inline ui-mini
                ui-corner-all ui-btn-b ui-btn-icon-left
                    ui-icon-phone">Contact Us</a>
    </h2>
</div>
</div><!-- /page -->
<!-- Catering page ends -->
```

Working with transitions and effects

One of the nice things with the jQuery Mobile framework is that we have control over the page transitions. Out of the box, jQuery Mobile supports nine (well, 10 if you count none) different transitions. These transitions are as follows:

- fade
- pop
- flip
- turn
- flow
- slidefade
- slide
- slideup
- slidedown
- none

These transitions are supported for both pages and dialogs. Using one is extremely easy. Simply declare it in your link declaration. Open up index.html and change your link for **About Us** to look like the following code:

```
<li><a href="#about" data-transition="flip">About Us</a></li>
```

If you want, go ahead and do that across all the pages, but at least do it for index.html right now.

Save the file and load up the project. Click on **About Us** and this time, instead of the basic fade, you will see the screen flip. Pretty cool, huh?

Go ahead and play around with other transitions and get a feel for how they work and find your favorite one. Once you find one you like, go ahead and set it for the rest of the links.

There are a couple things you should be aware regarding transitions. First off, unlike the biggest part of the framework, these transitions are not fully cross-platform. You may experience some odd behaviors, such as flickers or flashes on some devices. Luckily, the developers offer a fix for this, but it comes at a cost of efficiency on some browsers and Android devices. To fix this, in your custom CSS file, add the following code:

```
.ui-page { -webkit-backface-visibility: hidden; }
```

Another thing you may notice, particularly on Android devices running 2.x (which still counts for about 15 percent of the Android devices in use), is that the only transition you see is `fade`. This is because the browsers on that device do not have support for the 3D transformations. Android 3.x supports them; however, there still are some performance issues, so you may experience the aforementioned flickers on those devices.

The default transition is `fade` for the framework. If you don't specify one, like we did in the preceding code, each transition will be a `fade` one. You can change the default to whatever you like through setting the `defaultPageTransition` global option to whichever transition you want. You can also set the default for dialogs with the `defaultDialogTransition` global option.

Navigating with swipe gestures

Smartphones have really rewired the way most users navigate or want to navigate a website. Within native applications, most of the time we can swipe left, right, up, or down to get to the information we want. Occasionally, we may slip up and try this on websites that we view on our phones, and more times than not, we are met with a failure and remain on the same page that we were on when we started swiping.

Luckily for you and your future visitors and users, jQuery Mobile supports swiping gestures for left and right. We can capitalize on that and use it for navigation between our pages. It would be more difficult to do if we had our pages in individual files, as we previously did but, since we combined them all into one multi-page document, we can iterate through each of the pages in the file via swiping.

To implement this, we need to create a JavaScript function at the bottom of our current `index.html` file, right before the ending `body` tag. It can technically go where you like, but in the event of a JavaScript error, at least the rest of the page will load, whereas if the code is at the beginning and errors out, the page will be stuck.

Here is the function:

```
<script>
    $(body).bind("swipeleft swiperight",function(event)
    {
        var curPage = $.mobile.activePage[0];
        var direction = event.type;
        if(curPage.id == "home" && direction == "swipeleft")
            $.mobile.changePage("#about");
        if(curPage.id == "about" && direction == "swiperight")
            $.mobile.changePage("#home");
    });
</script>
```

For this test, we are only swiping between the Home and About pages. For the final product, we will go through all the pages.

Save the file and load this into a browser on one of your mobile devices (or a simulator that can simulate swipes), and swipe to the left, and then to the right.

Let's take a look at the code now:

```
$(body).bind("swipeleft swiperight",function(event)
```

Here, we start listening for the swipeleft and swiperight jQuery Mobile events.

```
var curPage = $.mobile.activePage[0];
var direction = event.type;
```

Next, we are assigning the current page to a variable, along with which event was used to another variable:

```
if(curPage.id == "home" && direction == "swipeleft")
    $.mobile.changePage("#about");
if(curPage.id == "about" && direction == "swiperight")
    $.mobile.changePage("#home");
```

Now, we are performing some if statements to check what the current page is and the direction of the swipe, depending on which we load the appropriate page.

Working with page loaders

Page loaders are a very nice piece of jQuery Mobile that allow you to let the users know that something is going on with your app, and that it is just not sitting idle or has not frozen.

Adding a page loader to your application is extremely easy. For our example, we are going to add one to our Events popup to let the user know we are loading events. For now, we will use the default icon, but in a later chapter, we will create our own so that it is customized.

Add the following code to the `<head>` section of your page:

```
<script>
    $(document).on("click", ".show-page-loading-msg",
        function() {
        var $this = $(this),
        theme = $this.jqmData("theme") ||
            $.mobile.loader.prototype.options.theme,
        msgText = "Loading Events",
            textVisible = $this.jqmData("textvisible") ||
                $.mobile.loader.prototype.options.textVisible,
            textonly = !!$this.jqmData("textonly"),
            html = $this.jqmData("html") || "";
        $.mobile.loading("show", {
            text: msgText,
            textVisible: textVisible,
            theme: theme,
            textonly: textonly,
            html: html
        });
        setTimeout(function() {
            $.mobile.loading("hide");
        }, 1000);
    });
</script>
```

Now, change our `Events` link (on all of the pages if you'd like, but definitely at least do it on the `home` div element) to the following code:

```
<a href="#eventregister" data-rel="popup"
    data-position-to="window" data-transition="pop"
        class="ui-btn-right ui-btn-b ui-btn ui-btn-inline ui-mini
            ui-corner-all ui-btn-icon-right ui-icon-calendar
                show-page-loading-msg" data-textonly="false"
                    data-textvisible="true" data-msgtext=""
                        data-inline="true">Events</a>
```

That's quite a long link, isn't it? Let's see what new we have added here. We just added the `show-page-loading-msg` class to the existing classes, and then also a variety of data attributes such as `data-textonly="false"`, `data-textvisible="true"`, `data-msgtext=""`, and `data-inline="true"`. Save it and load up in a browser. Now, when you click on the **Events** button, you will see our loading message with the spinning wheel, as seen in the following screenshot:

We have the `loader` widget in place, but as you must have already noticed, there is a small issue there. The popup has already been displayed, but we can still see the `loader` widget. We need to fix this. The `loader` widget should disappear as soon as the popup makes an appearance. This fix is pretty simple and straightforward:

1. First, add the following code in your `setTimeout` function:

   ```
   $("#eventregister").popup("open");
   ```

   Your `setTimeout` function will now look like the following code:

   ```
   setTimeout(function() {
       $.mobile.loading("hide");
       $("#eventregister").popup("open");
   }, 1000);
   ```

Here, we are basically triggering the opening of the popup using JavaScript and that too only after the `loader` widget has been hidden

2. Since we are opening the popup by using JavaScript, we need to remove the `data-rel="popup"` attribute from the `Events` button link in the header. The `Events` button link will now look like the following code:

```
<a href="#eventregister" data-position-to="window"
    data-transition="pop" class="ui-btn-right ui-btn-b
        ui-btn ui-btn-inline ui-mini ui-corner-all
            ui-btn-icon-right ui-icon-calendar
                show-page-loading-msg"
                data-textonly="false"
                    data-textvisible="true"
                        data-msgtext=""
                            data-inline="true">Events
                                </a>
```

Having made this change, the popup will now appear only after the loader message disappears. The output will now appear as seen in the following screenshot:

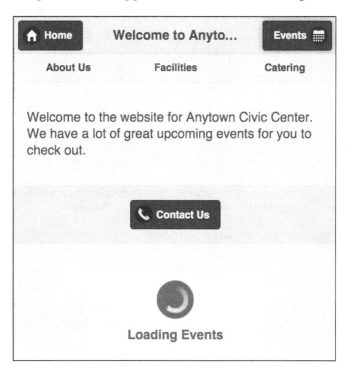

Let's break down the code now:

```
<script>
    $(document).on("click", ".show-page-loading-msg", function() {
        var $this = $(this),
        theme = $this.jqmData("theme") ||
            $.mobile.loader.prototype.options.theme,
        msgText = "Loading Events",
        textVisible = $this.jqmData("textvisible") ||
            $.mobile.loader.prototype.options.textVisible,
        textonly = !!$this.jqmData("textonly"),
        html = $this.jqmData("html") || "";
        $.mobile.loading("show", {
            text: msgText,
            textVisible: textVisible,
            theme: theme,
            textonly: textonly,
            html: html
        });
        setTimeout(function() {
            $.mobile.loading("hide");
            $("#eventregister").popup("open");
        }, 1000);
    });
</script>
```

What we are doing here is creating a global template, so to speak. We create the `loader` widget with all the required parameters that jQuery Mobile wants, and then at the end, we tell it to go away after 1,000 milliseconds. You can adjust this value as needed:

```
<a href="#eventregister" data-position-to="window"
    data-transition="pop" class="ui-btn-right ui-btn-b ui-btn
        ui-btn-inline ui-mini ui-corner-all ui-btn-icon-right
            ui-icon-calendar show-page-loading-msg"
                data-textonly="false" data-textvisible="true"
                    data-msgtext="" data-inline="true">Events</a>
```

Here, we are creating a loader with values specific to this instance. We could have simply created the `loader` widget with these values, but then all loaders would share those values. By doing it the way we did, we have much greater flexibility for our application.

Summary

We covered a lot of things in this chapter and this too was code-intense. We looked at the different modes of navigation that can be utilized within a jQuery Mobile framework and how these can be leveraged to give your end users the best experience on your mobile site.

We also explored how we can set up the navigation through the pages in your application by using swipe gestures. We explored various transitions and you now understand how these can be configured locally for individual pages or popups or even globally. And finally, the biggest thing that we achieved in this chapter is that we moved the project over to being a multipage document, instead of having multiple pages to maintain. We hope that you have followed the project so far and will also enjoy future chapters in which we will explore several different widgets and take our project forward.

In the next chapter, we will explore a few new widgets provided by the framework that will help us take our application to the next level. We will also take a look at how the jQuery Mobile framework interacts with third-party plugins. We will also cover the very important concept of accessibility by the end of the next chapter.

6

Controls and Widgets

In the previous chapter, we took our application from merely being a collection of pages to a fully structured web application bound using different techniques of navigation. We also looked at how we can leverage the touch capabilities of smartphones and provide users with a better mode of navigation using the swipe gesture. We concluded the previous chapter by creating a multi-page application for our project, the final working code for which you should have with you at this point.

Overview

The jQuery Mobile Framework comes with several out-of-the-box widgets, which are excellent and can be incorporated in our project very easily. In this chapter, we will take our Civic Center application to the next level and in the process of doing so, we will explore different widgets. We will explore the touch events provided by the jQuery Mobile framework further and then take a look at how this framework interacts with third-party plugins. We will be covering the following different widgets and topics in this chapter:

- Collapsible widget
- Listview widget
- Range slider widget
- Radio button widget
- Touch events
- Third-party plugins
- HammerJs
- FastClick
- Accessibility

Once again, we will continue working on our Civic Center project, so make sure you have the most updated code from the previous chapter. Now that we know what we will be completing as part of this chapter, grab your coffee, put on your developer hat, and get ready to begin a code-rich journey.

Widgets

We will not be using widgets for the first time in this project, but we will be looking at widgets in great detail in this chapter. We already made use of widgets as part of the Civic Center application. "Which? Where? When did that happen? What did I miss?" Don't panic as you have missed nothing at all. All the components that we use as part of the jQuery Mobile framework are widgets. The page, buttons, and toolbars are all widgets. So what do we understand about widgets from their usage so far?

One thing is pretty evident, widgets are feature-rich and they have a lot of things that are customizable and that can be tweaked as per the requirements of the design. These customizable things are pretty much the methods and events that these small plugins offer to the developers. So all in all:

> *Widgets are feature rich, stateful plugins that have a complete lifecycle, along with methods and events.*

We will now explore a few widgets as discussed before and we will start off with the collapsible widget. A collapsible widget, more popularly known as the accordion control, is used to display and style a cluster of related content together to be easily accessible to the user. Let's see this collapsible widget in action.

Pull up the index.html file that we have from the previous chapter. We will be adding the collapsible widget to the facilities page. You can jump directly to the content div of the facilities page. We will replace the simple-looking, unordered list and add the collapsible widget in its place. Add the following code in place of the `<ul>...<li></li>...</ul>` portion:

```
<div data-role="collapsibleset">
<div data-role="collapsible">
<h3>Banquet Halls</h3>
<p>List of banquet halls will go here</p>
</div>
<div data-role="collapsible">
<h3>Sports Arena</h3>
<p>List of sports arenas will go here</p>
</div>
<div data-role="collapsible">
```

```
    <h3>Conference Rooms</h3>
<p>List of conference rooms will come here</p>
</div>
<div data-role="collapsible">
<h3>Ballrooms</h3>
<p>List of ballrooms will come here</p>
</div>
</div>
```

That was pretty simple. As you must have noticed, we are creating a group of collapsibles defined by `div` with `data-role="collapsibleset"`. Inside this `div`, we have multiple `div` elements each with `data-role` of `"collapsible"`. These data roles instruct the framework to style `div` as a collapsible. Let's break individual collapsibles further. Each collapsible `div` has to have a heading tag (h1-h6), which acts as the title for that collapsible.

This heading can be followed by any HTML structure that is required as per your application's design. In our application, we added a paragraph tag with some dummy text for now. We will soon be replacing this text with another widget— `listview`. Before we proceed to look at how we will be doing this, let's see what the facilities page is looking like right now:

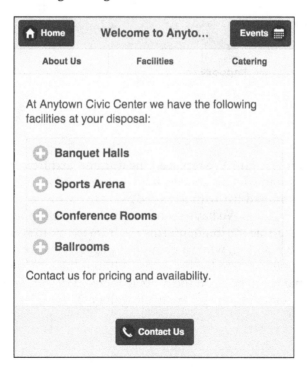

Now let's take a look at another widget that we will include in our project—the `listview` widget. The `listview` widget is a very important widget from the mobile website stand point. The `listview` widget is highly customizable and can play an important role in the navigation system of your web application as well.

In our application, we will include `listview` within the collapsible `div` elements that we have just created. Each collapsible will hold the relevant list items which can be linked to a detailed page for each item. Without further discussion, let's take a look at the following code. We have replaced the contents of the first collapsible list item within the paragraph tag with the code to include the `listview` widget. We will break up the code and discuss the minute details later:

```
    <div data-role="collapsible">
<h3>Banquet Halls</h3>
<p>
<span>We have 3 huge banquet halls named after 3 most celebrated
Chef's from across the world.</span>
<ul data-role="listview" data-inset="true">
<li>
<a href="#">Gordon Ramsay</a>
</li>
<li>
<a href="#">Anthony Bourdain</a>
</li>
<li>
<a href="#">Sanjeev Kapoor</a>
</li>
</ul>
</p>
    </div>
```

That was pretty simple, right? We replaced the dummy text from the paragraph tag with a span that has some details concerning what that collapsible list is about, and then we have an unordered list with `data-role="listview"` and some property called `data-inset="true"`. We have seen several data-roles before, and this one is no different. This data-role attribute informs the framework to style the unordered list, such as a tappable button, while a data-inset property informs the framework to apply the inset appearance to the list items. Without this property, the list items would stretch from edge to edge on the mobile device. Try setting the data-inset property to false or removing the property altogether. You will see the results for yourself.

Another thing worth noticing in the preceding code is that we have included an anchor tag within the `li` tags. This anchor tag informs the framework to add a right arrow icon on the extreme right of that list item. Again, this icon is customizable, along with its position and other styling attributes. Right now, our facilities page should appear as seen in the following image:

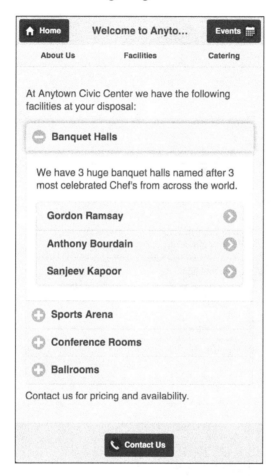

We will now add similar `listview` widgets within the remaining three collapsible items. The content for the next collapsible item titled **Sports Arena** should be as follows. Once added, this collapsible item, when expanded, should look as seen in the screenshot that follows the code:

```
<div data-role="collapsible">
    <h3>Sports Arena</h3>
    <p>
        <span>
```

```
              We have 3 huge sport arenas named after 3 most celebrated
   sport personalities from across the world.
          </span>
          <ul data-role="listview" data-inset="true">
              <li>
                  <a href="#">Sachin Tendulkar</a>
              </li>
              <li>
                  <a href="#">Roger Federer</a>
              </li>
              <li>
                  <a href="#">Usain Bolt</a>
              </li>
          </ul>
      </p>
  </div>
```

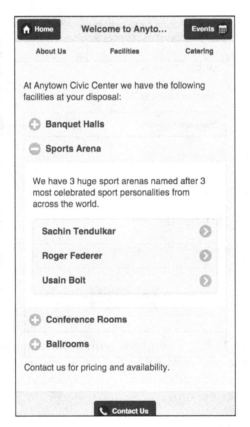

The code for the listview widgets that should be included in the next collapsible item titled **Conference Rooms**. Once added, this collapsible, item when expanded, should look as seen in the image that follows the code:

```
<div data-role="collapsible">
    <h3>Conference Rooms</h3>
    <p>
        <span>
            We have 3 huge conference rooms named after 3 largest
technology companies.
        </span>
        <ul data-role="listview" data-inset="true">
            <li>
                <a href="#">Google</a>
            </li>
            <li>
                <a href="#">Twitter</a>
            </li>
            <li>
                <a href="#">Facebook</a>
            </li>
        </ul>
    </p>
</div>
```

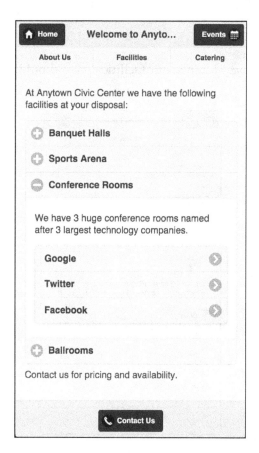

The final collapsible list item – **Ballrooms** – should hold the following code, to include its share of the `listview` items:

```
<div data-role="collapsible">
    <h3>Ballrooms</h3>
    <p>
        <span>
            We have 3 huge ball rooms named after 3 different dance
styles from across the world.
        </span>
        <ul data-role="listview" data-inset="true">
            <li>
                <a href="#">Ballet</a>
            </li>
            <li>
                <a href="#">Kathak</a>
            </li>
            <li>
                <a href="#">Paso Doble</a>
            </li>
        </ul>
    </p>
</div>
```

After adding these `listview` items, our facilities page should look as seen in the following image:

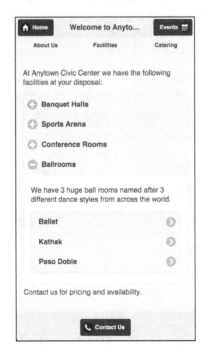

The `facilities` page now looks much better than it did earlier, and we now understand a couple more very important widgets available in jQuery Mobile – the collapsible widget and the `listview` Widget. We will now explore two form widgets – slider widget and the radio buttons widget.

For this, we will be enhancing our `catering` page. Let's build a simple tool that will help the visitors of this site estimate the food expense based on the number of guests and the type of cuisine that they choose. Let's get started then. First, we will add the required HTML, to include the slider widget and the radio buttons widget. Scroll down to the content `div` of the `catering` page, where we have the paragraph tag containing some text about the Civic Center's catering services.

Add the following code after the paragraph tag:

```
<form>
    <label style="font-weight: bold; padding: 15px 0px;"
for="slider">Number of guests</label>
    <input type="range" name="slider" id="slider" data-
highlight="true" min="50" max="1000" value="50">
    <fieldset data-role="controlgroup" id="cuisine-choices">
        <legend style="font-weight: bold; padding: 15px 0px;">Choose
your cuisine</legend>
        <input type="radio" name="cuisine-choice" id="cuisine-choice-
cont" value="15" checked="checked" />
        <label for="cuisine-choice-cont">Continental</label>
        <input type="radio" name="cuisine-choice" id="cuisine-choice-
mex" value="12" />
        <label for="cuisine-choice-mex">Mexican</label>
        <input type="radio" name="cuisine-choice" id="cuisine-choice-
ind" value="14" />
        <label for="cuisine-choice-ind">Indian</label>
    </fieldset>
    <p>
        The approximate cost will be: <span style="font-weight: bold;"
id="totalCost"></span>
    </p>
</form>
```

That is not much code, but we are adding and initializing two new form widgets here. Let's take a look at the code in detail:

```
<label style="font-weight: bold; padding: 15px 0px;"
for="slider">Number of guests</label>
<input type="range" name="slider" id="slider" data-highlight="true"
min="50" max="1000" value="50">
```

We are initializing our first form widget here—the slider widget. The slider widget is an input element of the type range, which accepts a minimum value and maximum value and a default value. We will be using this slider to accept the number of guests. Since the Civic Center can cater to a maximum of 1,000 people, we will set the maximum limit to 1,000 and we expect that we have at least 50 guests, so we set a minimum value of 50. Since the minimum number of guests that we cater for is 50, we set the input's default value to 50. We also set the data-highlight attribute value to true, which informs the framework that the selected area on the slider should be highlighted.

Next comes the group of radio buttons. The most important attribute to be considered here is the `data-role="controlgroup"` set on the `fieldset` element. Adding this data-role combines the radio buttons into one single group, which helps inform the user that one of the radio buttons is to be selected. This gives a visual indication to the user that one radio button out of the whole lot needs to be selected.

The values assigned to each of the radio inputs here indicate the cost per person for that particular cuisine. This value will help us calculate the final dollar value for the number of selected guests and the type of cuisine.

 Whenever you are using the form widgets, make sure you have the form elements in the hierarchy as required by the jQuery Mobile framework. When the elements are in the required hierarchy, the framework can apply the required styles.

At the end of the previous code snippet, we have a paragraph tag where we will populate the approximate cost of catering for the selected number of guests and the type of cuisine selected.

The catering page should now look as seen in the following image. Right now, we only have the HTML widgets in place. When you drag the slider or select different radio buttons, you will only see the UI interactions of these widgets and the UI treatments that the framework applies to these widgets. However, the total cost will not be populated yet. We will need to write some JavaScript logic to determine this value, and we will take a look at this in a minute. Before moving to the JavaScript part, make sure you have all the code that is needed:

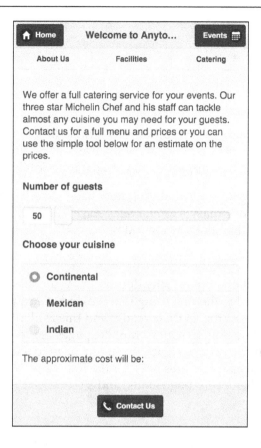

Now let's take a look at the magic part of the code (read JavaScript) that is going to make our widgets usable for the visitors of this Civic Center web application. Add the following JavaScript code in the script tag at the very end of our index.html file:

```
$(document).on('pagecontainershow', function(){
    var guests = 50;
    var cost = 35;
    var totalCost;
    $("#slider").on("slidestop", function(event, ui){
        guests = $('#slider').val();
        totalCost = costCal();
        $("#totalCost").text("$" + totalCost);
    });
    $("input:radio[name=cuisine-choice]").on("click", function() {
        cost = $(this).val();
        var totalCost = costCal();
```

```
        $("#totalCost").text("$" + totalCost);
    });
    function costCal(){
        return guests * cost;
    }
});
```

That is a pretty small chunk of code and pretty simple too. We will be looking at a few very important events that are part of the framework and that come in very handy when developing web applications with jQuery Mobile.

One of the most important things that you must have already noticed is that we are not making use of the customary `$(document).on('ready', function(){` in Jquery, but something that looks as the following code:

```
$(document).on('pagecontainershow', function(){
```

The million dollar question here is "why doesn't DOM already work in jQuery Mobile?" As part of jQuery, the first thing that we often learn to do is execute our jQuery code as soon as the DOM is ready, and this is identified using the `$(document).ready` function. In jQuery Mobile, pages are requested and injected into the same DOM as the user navigates from one page to another and so the DOM ready event is as useful as it executes only for the first page. Now we need an event that should execute when every page loads, and `$(document).pagecontainershow` is the one.

 The `pagecontainershow` element is triggered on the `toPage` after the transition animation has completed. The `pagecontainershow` element is triggered on the `pagecontainer` element and not on the actual page.

In the function, we initialize the guests and the cost variables to 50 and 35 respectively, as the minimum number of guests we can have is 50 and the "Continental" cuisine is selected by default, which has a value of 35. We will be calculating the estimated cost when the user changes the number of guests or selects a different radio button. This brings us to the next part of our code.

We need to get the value of the number of guests as soon as the user stops sliding the slider. jQuery Mobile provides us with the slidestop event for this very purpose. As soon as the user stops sliding, we get the value of the slider and then call the `costCal` function, which returns a value that is the number of guests multiplied by the cost of the selected cuisine per person. We then display this value in the paragraph at the bottom for the user to get an estimated cost.

We will discuss some more about the touch events that are available as part of the jQuery Mobile framework in the next section.

When the user selects a different radio button, we retrieve the value of the selected radio button, call the `costCal` function again, and update the value displayed in the paragraph at the bottom of our page.

If you have the code correct and your functions are all working fine, you should see something similar to the following image:

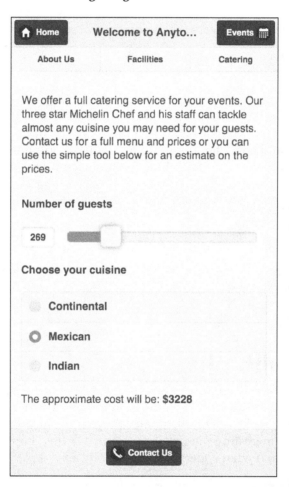

Input with touch

We discussed the `swipe`, `swiperight`, and `swipeleft` events in earlier chapters when we discussed navigating through the pages using touch. We will take a look at a couple of more touch events, which are `tap` and `taphold`. The `tap` event is triggered after a quick touch; whereas the `taphold` event is triggered after a sustained, long press touch.

The jQuery Mobile tap event is the gesture equivalent of the standard click event that is triggered on the release of the touch gesture. The following snippet of code should help you incorporate the `tap` event when you need to use it in your application:

```
$(".selector").on("tap", function(){
    console.log("tap event is triggered");
});
```

The jQuery Mobile `taphold` event triggers after a sustained, complete touch event, which is more commonly known as the long press event. The `taphold` event fires when the user taps and holds for a minimum of 750 milliseconds. You can also change the default value, but we will come to that in a minute. First, let's see how the `taphold` event is used:

```
$(".selector").on("taphold", function(){
    console.log("taphold event is triggered");
});
```

Now to change the default value for the long press event, we need to set the value for the following piece of code:

```
$.event.special.tap.tapholdThreshold
```

Working with plugins

A number of times, we will come across scenarios where the capabilities of the framework are just not sufficient for all the requirements of your project. In such scenarios, we have to make use of third-party plugins in our project. We will be looking at two very interesting plugins in the course of this chapter, but before that, you need to understand what jQuery plugins exactly are.

A jQuery plugin is simply a new method that has been used to extend jQuery's prototype object. When we include the jQuery plugin as part of our code, this new method becomes available for use within your application. When selecting jQuery plugins for your jQuery Mobile web application, make sure that the plugin is optimized for mobile devices and incorporates touch events as well, based on your requirements.

The first plugin that we are going to look at today is called **FastClick** and is developed by FT Labs. This is an open source plugin and so can be used as part of your application. FastClick is a simple, easy-to-use library designed to eliminate the 300 ms delay between a physical tap and the firing on the click event on mobile browsers.

Wait! What are we talking about? What is this 300 ms delay between tap and click? What exactly are we discussing? Sure. We understand the confusion. Let's explain this 300 ms delay issue.

The click events have a 300 ms delay on touch devices, which makes web applications feel laggy on a mobile device and doesn't give users a native-like feel. If you go to a site that isn't mobile-optimized, it starts zoomed out. You have to then either pinch and zoom or double tap some content so that it becomes readable. The double-tap is a performance killer, because with every tap we have to wait to see whether it might be a double tap—and this wait is 300 ms. Here is how it plays out:

1. `touchstart`
2. `touchend`
3. Wait 300ms in case of another tap
4. `click`

This pause of 300 ms applies to click events in JavaScript, but also other click-based interactions such as links and form controls. Most mobile web browsers out there have this 300 ms delay on the click events, but now a few modern browsers such as Chrome and FireFox for Android and iOS are removing this 300 ms delay. However, if you are supporting the older Android and iOS versions, with older mobile browsers, you might want to consider including the FastClick plugin in your application, which helps resolve this problem.

Let's take a look at how we can use this plugin in any web application. First, you need to download the plugin files, or clone their GitHub repository here: `https://github.com/ftlabs/fastclick`. Once you have done that, include a reference to the plugin's JavaScript file in your application:

```
<script type="application/javascript" src="path/fastclick.js"></script>
```

Make sure that the script is loaded prior to instantiating FastClick on any element of the page. FastClick recommends you to instantiate the plugin on the body element itself. We can do this using the following piece of code:

```
$(function){
    FastClick.attach(document.body);
}
```

That is it! Your application is now free of the 300 ms click delay issue and will work as smooth as a native application.

> We have just provided you with an introduction to the FastClick plugin. There are several more features that this plugin provides. Make sure you visit their website— `https://github.com/ftlabs/fastclick`—for more details on what the plugin has to offer.

Another important plugin that we will look at is **HammerJs**. HammerJs, again is an open source library that helps recognize gestures made by touch, mouse, and pointerEvents. Now, you would say that the jQuery Mobile framework already takes care of this, so why do we need a third-party plugin again? True, jQuery Mobile supports a variety of touch events such as tap, tap and hold, and swipe, as well as the regular mouse events, but what if in our application we want to make use of some touch gestures such as pan, pinch, rotate, and so on, which are not supported by jQuery Mobile by default? This is where HammerJs comes into the picture and plays nicely along with jQuery Mobile. Including HammerJS in your web application code is extremely simple and straightforward, like the FastClick plugin. You need to download the plugin files and then add a reference to the plugin JavaScript file:

```
<script type="application/javascript" src="path/hammer.js"></script>
```

Once you have included the plugin, you need to create a new instance on the `Hammer` object and then start using the plugin for all the touch gestures you need to support:

```
var hammerPan = new Hammer(element_name, options);
hammerPan.on('pan', function(){
    console.log("Inside Pan event");
});
```

By default, `Hammer` adds a set of events—tap, double tap, swipe, pan, press, pinch, and rotate. The pinch and rotate recognizers are disabled by default, but can be turned on as and when required.

> HammerJS offers a lot of features that you might want to explore. Make sure you visit their website—`http://hammerjs.github.io/` to understand the different features the library has to offer and how you can integrate this plugin within your existing or new jQuery Mobile projects.

Accessibility

Most of us today cannot imagine our lives without the Internet and our smartphones. Some will even argue that the Internet is the single largest revolutionary invention of all time that has touched numerous lives across the globe. Now, at the click of a mouse or the touch of your fingertip, the world is now at your disposal, provided you can use the mouse, see the screen, and hear the audio—impairments might make it difficult for people to access the Internet. This makes us wonder about how people with disabilities would use the Internet, their frustration in doing so, and the efforts that must be taken to make websites accessible to all.

Though estimates vary on this, most studies have revealed that about 15% of the world's population have some kind of disability. Not all of these people would have an issue with accessing the web, but let's assume 5% of these people would face a problem in accessing the web. This 5% is also a considerable amount of users, which cannot be ignored by businesses on the web, and efforts must be taken in the right direction to make the web accessible to these users with disabilities.

jQuery Mobile framework comes with built-in support for accessibility. jQuery Mobile is built with accessibility and universal access in mind. Any application that is built using jQuery Mobile is accessible via the screen reader as well. When you make use of the different jQuery Mobile widgets in your application, unknowingly you are also adding support for web accessibility into your application. jQuery Mobile framework adds all the necessary `aria` attributes to the elements in the DOM. Let's take a look at how the DOM looks for our facilities page:

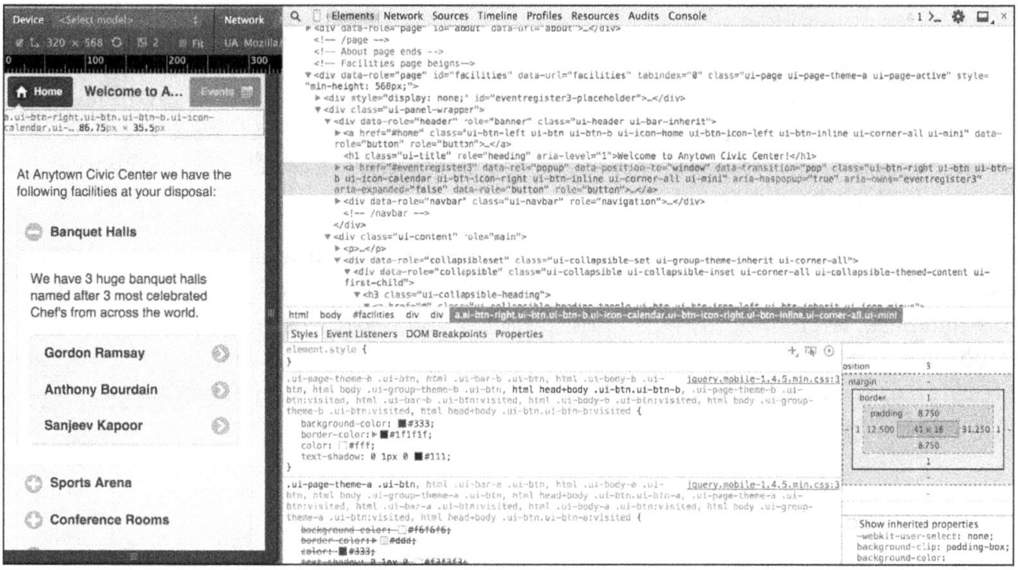

Look at the highlighted **Events** button in the top right corner and its corresponding HTML (also highlighted) in the developer tools. You will notice that there are a few attributes added to the anchor tag that start with `aria-`. We did not add any of these `aria-` attributes when we wrote the code for the **Events** button. jQuery Mobile library takes care of these things for you. The accessibility implementation is an ongoing process and the awesome developers at jQuery Mobile are working towards improving the support every new release.

We spoke about `aria-` attributes, but what do they really represent? **WAI - ARIA** stands for **Web Accessibility Initiative – Accessible Rich Internet Applications**. This was a technical specification published by the **World Wide Web Consortium (W3C)** and basically specifies how to increase the accessibility of web pages. ARIA specifies the roles, properties, and states of a web page that make it accessible to all users.

Accessibility is extremely vast, hence covering every detail of it is not possible. However, there is excellent material available on the Internet on this topic and we encourage you to read and understand this. Try to implement accessibility into your current or next project even if it is not based on jQuery Mobile. Web accessibility is an extremely important thing that should be considered, especially when you are building web applications that will be consumed by a huge consumer base—on e-commerce websites for example.

Summary

In this chapter, we made use of some of the available widgets from the jQuery Mobile framework and we built some interactivity into our existing Civic Center application. The widgets that we used included the range slider, the collapsible widget, the listview widget, and the radio button widget. We evaluated and looked at how to use two different third-party plugins—FastClick and HammerJs. We concluded the chapter by taking a look at the concept of Web Accessibility.

In the next chapter, we will rig our Civic Center application with some dynamic data using PHP to connect to a MySQL database. We will take a look at how to connect to a database, fetch data, and also post data to a database. The next chapter is going to be very interesting and code-intensive. Take a break, grab a coffee, and come back prepared to go on an exciting journey to make our application truly dynamic.

7
Working with Data

In the previous chapter, we added some really nice functionalities to our Civic Center application and took a step toward making our application truly interactive. In this chapter, we will take our application to the next level by giving it the capability of playing with dynamic data. We will connect our jQuery Mobile frontend code to a MySQL database to get and fetch data via PHP. Hope you are as excited as we are to add this dynamic functionality to our Civic Center application.

Overview

You now know what we are going to try to achieve in this chapter. We will assume that you have some basic understanding of PHP, and even if you have never worked with PHP before, that is fine. We are pretty sure that you will very easily be able to follow along with us to add the PHP functionality in our code. By the end of this chapter, if you feel you have gained some interest in PHP, we encourage you to take a look at one of the several titles published by Packt Publishing on this topic. In this chapter, we will cover the following points:

- Creating a MySQL database using MySQL in XAMPP setup
- Creating a table and adding some data to it
- Connecting to a MySQL database from PHP
- Fetching data from a table and displaying this data
- Forms and client-side validations
- Inserting data into a database table

Once again, make sure you have the latest and the most accurate code from the previous chapter as we will continue editing our Civic Center application. Wear your developer hats once again and let's rock 'n' roll.

Setting up our database

Remember, in *Chapter 1, Getting Started,* we had said that we will do a small bit of PHP and MySQL later, and so we had installed XAMPP and verified that it was running fine on our local system. Now is the time to make use of that installed XAMPP server for PHP and MySQL. We have to first start our **MySQL Database** instance on XAMPP. To do this, you need to open up your XAMPP console and go to the **Manage Servers** tab. You will notice that the **MySQL Database** server status is **Stopped**. We need to start this server:

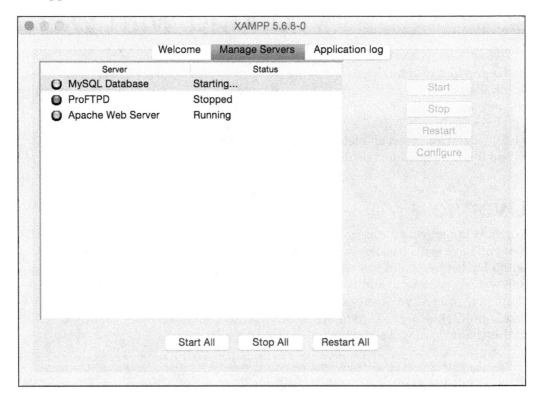

Starting your local instance of **MySQL Database** in XAMPP is the first step toward working on MySQL. Now, open up phpMyAdmin in your browser using the localhost/phpmyadmin URL. You might be asked for your login credentials. If you have not changed the default credentials, you should be able to login using the following credentials:

- **Username:** root

- **Password:** <No password. Blank field>

You should now see the **phpMyAdmin** screen as shown in the following image:

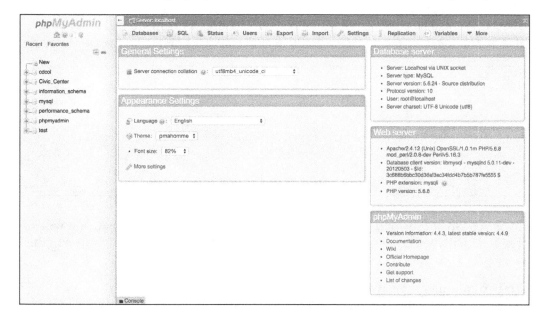

We will now create a new database for our Civic Center application, add a new table in it, and then add a few data rows in this table.

To create a new database, click on the **New** button at the top in the left navigation. You will be presented with a text field asking for the name of the database. Let's call our database `Civic_Center`. Click on the **Create** button:

As soon as you hit the **Create** button, you will be presented with a new view that asks to create a table. Let's create a table named Events_Catalog with two columns for now. We will be starting off with two columns, and later on we will modify this table to hold data for our events scheduled at the Civic Center:

Hit on the **Go** button on the bottom-right corner. As soon as you do that, you will be presented with yet another view asking for the column names and their types in this Event_Catalog table. Let's go ahead and fill this up.

Our two column names would be Event_ID and Event_Name of type INT and VARCHAR, respectively. The Event_ID column would be our primary key of length 2. The Event_Name field's length will be restricted to 50. Make all these changes; you can refer to the following screenshot for this:

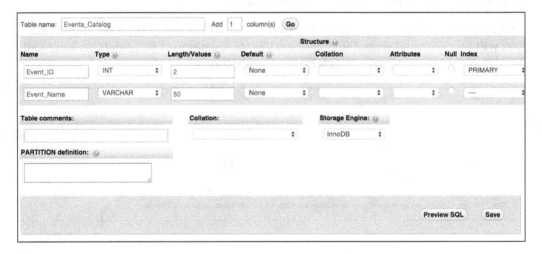

Hit the **Save** button and now you will see a new view that shows you the structure of the table that we just created. We now have the table created, but there is no data in it. A database table with no data rows is not very useful for us. So let's move ahead and add some events data into our table. To add data to the table, click on the **Insert** tab that you see at the top of your current **phpMyAdmin** window:

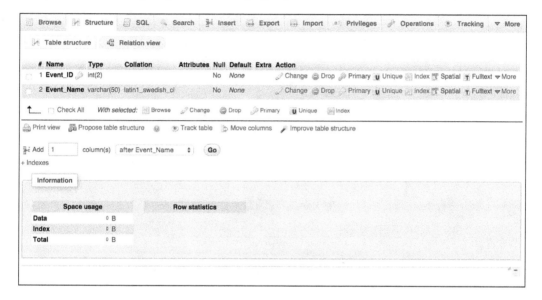

You will see a new form that asks for data values. By default, you will see fields for two data rows. Let's add five rows to our table, and to do that, select **5** in the select box at the very bottom of the page that says **Continue insertion with 2 rows**. Let's add the following data in these fields:

Event_ID	Event_Name
1	HTML5 Meetup
2	AnyTown Community Annual Awards
3	StepUp Dance Class
4	HTML5 Meetup Dinner
5	Certified Scrum Master Training

Your screen should now look as shown in the following image. Add all the data to the right fields and click on the **Go** button at the bottom:

That's it! We have our database table ready with all the data we need! This was pretty simple, and we hope you have followed us through and have all the right things in place. If you haven't got these right, just track back to what went wrong and complete the process. Getting the data in the right place is extremely critical for the future course of action in this chapter.

Now that we have the database table ready, we will move on to some code again, this time PHP and not jQuery Mobile. Let us see how to connect to this database table via PHP in the following section.

A dab of PHP

Before we start writing any code, let's set up a few simple things:

1. Let's create a new folder named JQMData in the htdocs folder in our XAMPP folder. Look for your XAMPP folder in your default installation drive on Windows or the Applications directory on Mac, if you have your XAMPP installed at the default location.

2. Now, copy the index.html file from the previous chapter and paste it into this JQMData folder.

3. Rename this file from index.html to index.php.

4. Create a new file and name it db.php.

If you have followed the steps correctly, you should see our application when you fire up this URL in your browser: localhost/JQMData.

We are now ready to write our first PHP code for this application. Open up the db.php file that we have just created. We have to now establish a connection to the MySQL database that we have created. Let's see how this works. Copy the following code to db.php file and save it:

```php
<?php
  $servername = "localhost";
  $username = "root";
  $password = "";
  $dbname = "Civic_Center";

  // Create connection
  $conn = mysqli_connect($servername, $username, $password, $dbname);
  if (!$conn) {
    die("Connection failed: " . mysqli_connect_error());
  }
  else{
    echo "Connection successful!";
  }
  mysqli_close($conn);
?>
```

The contents of the file are very simple and straightforward to understand. To connect to a database, we need to set the authentication parameters, which are basically the name of your server, the name of the database you wish to connect to, and the login credentials to access this database. The first four lines of code do exactly this:

```php
$servername = "localhost";
$username = "root";
$password = "";
$dbname = "Civic_Center";
```

We declared four variables and assigned the right values to each one of them. Now we have to make an attempt to connect to this database. To do that, we have the next line of code:

```
mysqli_connect($servername, $username, $password, $dbname);
```

The `mysqli_connect` method is a method that accepts four parameters, namely, the name of the server to connect to, the name of the database to connect to, the username, and the password. When this method executes, we either get a success or a failure response. We have saved this response in the `$conn` variable.

Next comes the if-else block, where we do some stuff based on whether the connection was established without any errors. The happy path scenario here would be that the execution of the code goes into the else block, instead of the if block! The last line basically closes the connection that we have established with the database. Always remember to close the connection to the database once all your data communication activities are complete.

Now let's see if we get a success message on our web browser. To verify that, fire up your web browser and access this URL: `http://localhost/JQMData/db.php`.

You should see the **Connection successful!** message being displayed on your screen. Hooray! we have successfully created a database table with data, and now we are able to establish connection with it via PHP. In the next section, we will take a look at how we can retrieve the data from the database table.

Fetching data

Once the connection with a MySQL database is successful, the next logical step is to fetch the data from this database table. In this section, we will see how we would fetch the data from a database table once we are connected to it.

Let us continue editing the `db.php` file. We will add the following code in the else block where we were just displaying a successful message. Displaying a success message is not very meaningful, and so we will replace that `echo` statement with the following code:

```
$sql = "SELECT Event_ID, Event_Name FROM Events_Catalog";
$result = mysqli_query($conn, $sql);

if (mysqli_num_rows($result) > 0) {
    // output data of each row
    while($row = mysqli_fetch_assoc($result)) {
```

```
        echo "Event ID is: <b>" . $row["Event_ID"] . "</b><br /> Event
name is: <b>" . $row["Event_Name"] . "</b><br /><br />";
    }
}
else {
    echo "<h3>Sorry, currently there are no scheduled events.</h3>";
}
```

Again, as you must have observed, a small piece of code is going to help us fetch data from a database table. The first line of code is a standard SQL query to select data from two columns Event_ID and Event_Name from our Events_Catalog table. The next line of code is a PHP method that passes the SQL query to the established connection with the database.

What follows is a standard if-else block, where we check if there is any data that we have received from the SQL query. If there is one or more rows, then we will run a while loop to retrieve all the rows from the Events_Catalog table. When we run this while loop, we are displaying each row on the browser using the echo statement and some very basic styling. In case there was no data in the Events_Catalog table, meaning there were no rows at all, the code will go into the else block, where it will display a message to the user informing him/her that there are no rows to display. Always make sure that you take care of error or negative scenarios and handle them gracefully.

Fire up the dp.php page on your web browser like we did before, and you should see all the data we had inserted in the Events_Catalog table in the following format:

Event ID is: **1**
Event name is: **HTML5 Meetup**

Event ID is: **2**
Event name is: **AnyTown Community Annual Awards**

Event ID is: **3**
Event name is: **StepUp Dance Class**

Event ID is: **4**
Event name is: **HTML5 Meetup Dinner**

Event ID is: **5**
Event name is: **Certified Scrum Master Training**

That was simple, wasn't it? Now we know how to connect to a database table via PHP and also how to fetch data from it. But we can't just display the data that we received from a database table on a web page and leave it at that. The data should also be presented in a meaningful manner so that the end-user can interact with it. We will take a look at how we can use the logic so far in our jQuery Mobile code, and change the face of the application. We will do this in the following section.

Displaying information

Open up your `index.php` file now, along with the `db.php` file. We will be making changes to these files in order to make the code easier to manage and maintain. All the events data that we have created and are able to fetch from the database table is to be displayed in the **Events** popup, which till this moment has been displaying the **Sorry, currently there are no scheduled events.** message, as shown in the following screenshot:

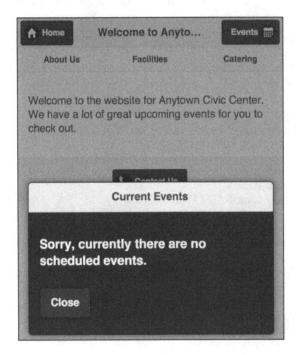

To do this, we will be modifying the following piece of code on all the pages that we have created, which are **Home**, **About Us**, **Facilities**, and **Catering**:

```
<div role="main" class="ui-content">
    <h3 class="ui-title">Sorry, currently there are no scheduled
events.</h3>
```

```
    <a href="#" class="ui-btn ui-corner-all ui-shadow ui-btn-inline
ui-btn-b" data-rel="back">Close</a>
    </div>
```

Instead of displaying this static message, we will now show some dynamic data here. In order to display a list of events, we will make use of the good old `listview` widget. We will also include the data fetch logic from the `db.php` file here so that the code becomes manageable, and we will separate out the configuration of database logic from the presentation layer.

Remove the following code from the `db.php` file:

```php
if (!$conn) {
  die("Connection failed: " . mysqli_connect_error());
}
else{
  $sql = "SELECT Event_ID, Event_Name FROM Events_Catalog";
    $result = mysqli_query($conn, $sql);

    if (mysqli_num_rows($result) > 0) {
        // output data of each row
        while($row = mysqli_fetch_assoc($result)) {
            echo "Event ID is: <b>" . $row["Event_ID"] . "</b><br />
Event name is: <b>" . $row["Event_Name"] . "</b><br /><br />";
        }
    }
    else {
        echo "<h3>Sorry, currently there are no scheduled events.</
h3>";
    }
}
mysqli_close($conn);
```

Add the preceding code inside the `div` element with `role="main"` in place of the `<h3>` tag. We have to retain the **Close** button as it is. Make sure you remove the last line of code from the preceding block. Remember to add this same code on all pages, and all of these are in one single PHP file—the `index.php` file. If we close the connection to the database on the **Home** page itself, we will get PHP database connection failure errors on the rest of the pages. So, now your popup's content **div** with `role="main"` should have the following code:

```php
<?php
<div role="main" class="ui-content">
if (!$conn) {
  die("Connection failed: " . mysqli_connect_error());
```

```
    }
    else{
      $sql = "SELECT Event_ID, Event_Name FROM Events_Catalog";
        $result = mysqli_query($conn, $sql);

        if (mysqli_num_rows($result) > 0) {
            // output data of each row
            while($row = mysqli_fetch_assoc($result)) {
                echo "Event ID is: <b>" . $row["Event_ID"] . "</b><br />
Event name is: <b>" . $row["Event_Name"] . "</b><br /><br />";
            }
        }
        else {
            echo "<h3>Sorry, currently there are no scheduled events.</
h3>";
        }
    }
    <a href="#" class="ui-btn ui-corner-all ui-shadow ui-btn-inline ui-
btn-b" data-rel="back">Close</a>
    </div>
    ?>
```

Make sure you add the PHP `<?php ?>` statements around the preceding PHP code, so that the web server understands that it is a PHP block and executes the logic within. Another thing that we have to do is style this data that we have fetched as list items. In order to do that, we will just modify the `while` block from the code. Copy the following piece of code in place of the `while` block:

```
while($row = mysqli_fetch_assoc($result)) {
    $Event_ID = $row['Event_ID'];
    echo "<li><a href='#'>" . $row["Event_Name"] . "</a></li>";
}
```

You will notice that we have added the `<li>` element for every data row and also have an `anchor` tag within it, so that the jQuery Mobile framework can style it as a `listview` widget with a default right arrow on every list item. We have added the `<li>` elements, but there is no `<ul>` element with `data-role="listview"` around this. We will add that outside the PHP block. Make sure you add the following `<ul>` tag just above the `<?php` line and just after `div` with `data-role="main"` line of code:

```
<ul data-role="listview" data-inset="true" data-theme="a">
```

You will close the `<ul>` tag after the `?>` line of code, where the PHP block ends. To avoid any confusion and for your reference, here is the entire block of code:

```
<div role="main" class="ui-content">
    <ul data-role="listview" data-inset="true" data-theme="a">
        <?php
            // Check connection
            if (!$conn) {
                die("Connection failed: " . mysqli_connect_error());
            }
            $sql = "SELECT Event_ID, Event_Name FROM Events_Catalog";
            $result = mysqli_query($conn, $sql);

            if (mysqli_num_rows($result) > 0) {
                // output data of each row
                while($row = mysqli_fetch_assoc($result)) {
                    $Event_ID = $row['Event_ID'];
                    echo "<li><a href='#'>" . $row["Event_Name"] . "</a></li>";
                }
            }
            else {
                echo "<h3 class='ui-title'>Sorry, currently there are no scheduled events.</h3>";
            }
        ?>
    </ul>
    <a href="#" class="ui-btn ui-corner-all ui-shadow ui-btn-inline ui-btn-b" data-rel="back">Close</a>
</div>
```

Make sure you have all the changes as per the previous code and on all the pages— **Home**, **About Us**, **Facilities**, and **Catering**. There is one final change that we have to make to get this all to work together. We removed the data fetch logic from the `db.php` file, but the database connectivity logic still sits there:

```
<?php
    $servername = "localhost";
    $username = "root";
    $password = "";
    $dbname = "Civic_Center";

    // Create connection
    $conn = mysqli_connect($servername, $username, $password, $dbname);
?>
```

We need this logic to fire before we are able to display anything in our events popup. We can do this with just a single line of code. Add the following piece of code just after the opening of the body tag in the index.php file:

```php
<?php
    include 'db.php';
?>
```

The include keyword basically fetches the contents of the db.php file into our current file, which is index.php. Once you have done this change, it's time to see our code in action. Fire up the following URL in your web browser and you should see our Civic Center application. Now, click on the **Events** button in the top-right corner and you should be able to see a list of events in a popup: http://localhost/JQMData/index.php:

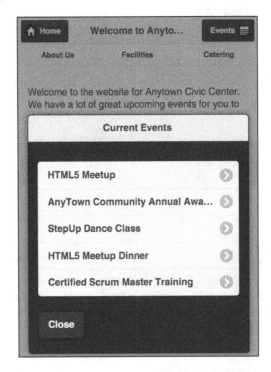

Isn't this impressive? Doesn't this look pretty cool? With very minimal PHP code, we have been able to add dynamics to our otherwise static application. Right now, if you click on any of these list items, nothing will happen. But we will add some functionality to it in the next section. The user would be able to navigate to a new page that shows the details of the event that he has clicked/tapped on. However, before we do that, we will have to make some additions to our database table to store some more information, some more columns. Let's head to it straight away.

Some more PHP

Open up the `Events_Catalog` table details in your **phpMyAdmin**. Click on the **Structure** tab. Toward the bottom of the page, you will notice a text field to add more columns to the existing table. We will be adding three more columns to the existing two columns. Make the required changes and click on the **Go** button toward the right end of the same row:

You will now be presented with fields, where you need to fill in the necessary details. The three new columns that we are going to create are `Event_Organizer`, `Event_Loc`, and `Event_Desc`. The field type and field length details are shown in the following image. Make the necessary changes and click on the **Save** button at the bottom of the screen:

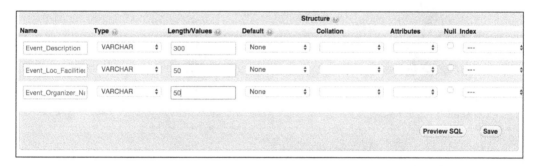

Now select all five rows and click on the **Change** link. You will be presented with a screen that shows all the columns for every row as input fields, which can be updated or changed as per our needs. We will make the following changes:

Event_ID	Event_Name	Event_Description	Event_Loc_Facilities	Event_Organizer_Name
1	HTML5 Meetup	HTML5 Meetup related description will come here	Google Conference Room	HTML5Fans Meetup Group
2	AnyTown Community Annual Awards	Some description will come here	Sachin Tendulkar Sports Arena	AnyTown Cultural Committee

Event_ID	Event_Name	Event_Description	Event_Loc_Facilities	Event_Organizer_Name
3	StepUp Dance Class	Some description will come here	Ballet Ballroom	StepUp Dance Training Institute
4	HTML5 Meetup Dinner	Some description will come here	Gordon Ramsay Banquet Hall	HTML5Fans Meetup Group
5	Certified Scrum Master Training	Some description will come here	Facebook Conference Room	Global Training Institute

The screen should be as shown in the following image:

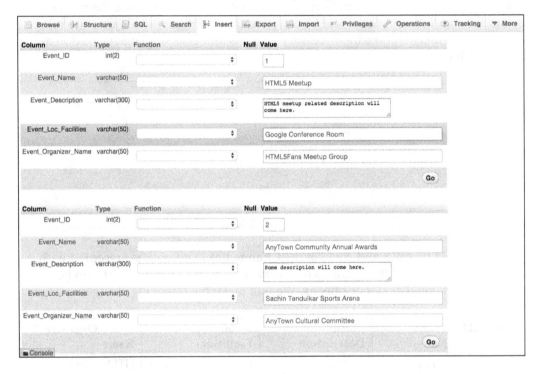

Now that we have our database table ready for the new functionality, let's turn our attention to some more PHP code. Let us look at this one step at a time:

1. Let's start by adding a value to the href attribute of the anchor tag within the list items in the preceding code. The list item, `<li>` tag, is within the PHP `while` block:

   ```
   while($row = mysqli_fetch_assoc($result)) {
       $Event_ID = $row['Event_ID'];
       echo "<li><a href='#'>" . $row["Event_Name"] . "</a></li>";
   }
   ```

   Add the following in place of the # character preceding the href attribute:

   ```
   href='event.php?id=$Event_ID'
   ```

   Make sure you make this change at all four places (in all pages) inside the index.php file. What we are doing here is basically sending a GET request to the event.php page with the ID of the event that is currently selected by the user.

2. As you must have noticed, we are redirecting to an event.php page, but we don't have one yet. No problem. Let's create a new event.php file. This is going to be our new and the fifth jQuery Mobile page of this project. It will have the exact same structure as that of any other page. In order to do that, we will replicate the HTML from some other page to this. Let's copy the HTML of the **Home** page, which is the div element with data-role="page" and id="home", into the event.php file. Make sure you add this div page inside a body tag.

3. Also copy the following code from index.php and add it just after the opening of the body tag:

   ```
   <?php
   include 'db.php';
   ?>
   ```

4. We have copied the div page but we need the head section of index.php as well. This holds our references to the jQuery library, jQuery Mobile JS, and CSS, and the script to show the events popup. Make sure when you copy the contents of the `<head>` tag into event.php, you also add the HTML tag at the top and close it in the very end.

5. Save this event.php file.

6. Let's make small changes to the IDs of different things, so that they don't conflict with other elements having same ID. Replace id="home" with id="event" on the div page with data-role="page".

7. On the next line that follows, replace id="eventregister" with id="eventregister5" on div with data-role="popup".

8. Scroll down to the div element with data-role="header". A couple of lines after this, on the anchor tag, replace href="#eventregister" by href="#eventregister5".

9. Now, scroll further down to the main content of the page, to the div page with role="main". You will notice the following code:

```
<p>Welcome to the website for Anytown Civic Center. We have a lot
of great upcoming events for you to check out.</p>
```

We will be replacing the text in this paragraph with the following PHP code:

```php
<?php
    // Check connection
    if (!$conn) {
        die("Connection failed: " . mysqli_connect_error());
    }
    $event_id = $_GET['id'];
    $sql = "SELECT Event_Name, Event_Description, Event_Loc_
Facilities, Event_Organizer_Name FROM Events_Catalog WHERE Event_
ID=$event_id";
    $result = mysqli_query($conn, $sql);
    if (mysqli_num_rows($result) > 0) {
        // output data of each row
        $row = mysqli_fetch_assoc($result);
            echo "<b>Event: </b><br />" . $row['Event_Name'] .
"<br /><br />" .
                "<b>Organized By: </b><br />" . $row['Event_
Organizer_Name'] . "<br /><br />" .
                "<b>Event Location: </b><br />" . $row['Event_Loc_
Facilities'] . "<br /><br />" .
                "<b>Brief Description: </b><br />" . $row['Event_
Description'];
    }
    else {
        echo "<h3 class='ui-title'>Sorry, event details are
unavailable.</h3>";
    }
    mysqli_close($conn);
?>
```

`$_GET['id']` will retrieve the event ID of the event that the user has clicked on, and that is the one that is passed in the URL. Based on this event ID, we will construct an SQL query to retrieve the remaining details related to that event ID. The SQL query that we construct will look like the following:

```
"SELECT Event_Name, Event_Description, Event_Loc_Facilities,
Event_Organizer_Name FROM Events_Catalog WHERE Event_ID=$event_
id";
```

This SQL query will always return only one row, hence we do not need a `while` loop to loop through the result. We then apply some style to the fetched data and display the same.

10. Save the file and head straight to your browser to see these changes in action. You should be able to click on any of the events from the list of events in the popup, and when you click on one, you should be redirected to a page, as seen in the following, whose URL has the following format: `http://localhost/JQMData/event.php?id=2`:

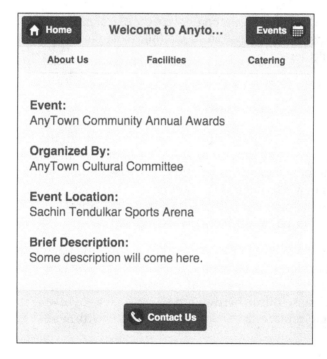

Forms and validation

So far we have only created a database table and fetched the data from it. We haven't seen how we can post data to a database table. Over the course of the next couple of sections, we will see how we can post the data from the frontend to the database via PHP. In order to do that, we will create a form that accepts some values and we will post these values to the database. However, before posting these values, we will validate these form fields on the frontend.

Let's consider a scenario where a user of the Civic Center application wants to make a reservation for an event he/she wishes to host at one of the facilities the Civic Center provides. We will create a form that asks the user for some details, so that when the user submits these, the authorities at the Civic Center can get in touch with the customer. We will create a new page inside the `index.php` file with `id="contact"` and the user will be able to access this page from the **Home** page, where we will add a link to the **Contact** page.

Now that we have a clear picture of what we want to achieve, let's start working toward it. First, we will modify the contents of the `div` element with `data-role="main"` for the `div` page with `id="home"`. Add the following paragraph to the existing one here:

```
<p>If you wish to make reservations to avail a facility at the Civic
Center, please fill out the <a href="#contact">contact form</a>, and
we will reach you!</p>
```

Now let's create a new `div` page with `id="contact"` within the `index.php` file. For this purpose, let's copy the `div` page with `id="home"` and we will make changes to the IDs of different elements as we did in the case of our **Events** page earlier. We will add this new `div` page at the bottom of the file just before the script that we included right before closing the `<body>` tag. Our **Home** page has become like a template for all other pages we have in this application.

Now, we need to make the changes to the IDs in this new page so that there are no conflicts with other elements. First, replace the `div` page `id="home"` with `id="contact"`. On the very next line, replace `id="eventregister"` with `id="eventregister6"` on `div` with `data-role="popup"`. Now, scroll further down to `div` with `data-role="header"`. Just a couple of lines, on the `<a>` tag, replace `href="#eventregister"` with `href="#eventregister6"`.

We will now replace the content of this page and the contents of `div` with `role="main"`, and add the following form code in place of the paragraph content that we have right now:

```html
<form class="user" novalidate>
    <label for="name">Name:</label>
    <input type="text" name="name" id="name" value="" required
maxlength="20" />
    <div style="color: red; font-size: 13px;"></div>
    <label for="email">Email Id:</label>
    <input type="email" name="email" id="email" value="" required />
    <div style="color: red; font-size: 13px;"></div>
    <label for="contact">Contact Number:</label>
    <input type="number" name="contact" id="contact" value="" required
/>
    <div style="color: red; font-size: 13px;"></div>
    <label for="facility">Select a facility:</label>
    <select name="facility" id="facility" value="" required>
        <optgroup label="Banquet Halls">
            <option value="GR">Gordan Ramsay</option>
            <option value="AB">Anthony Bourdain</option>
            <option value="SK">Sanjeev Kapoor</option>
        </optgroup>
        <optgroup label="Sports Arena">
            <option value="ST">Sachin Tendulkar</option>
            <option value="RF">Rager Federrer</option>
            <option value="UB">Usain Bolt</option>
        </optgroup>
        <optgroup label="Conference Rooms">
            <option value="G">Google</option>
            <option value="T">Twitter</option>
            <option value="F">Facebook</option>
        </optgroup>
        <optgroup label="Ball Rooms">
            <option value="B">Ballet</option>
            <option value="PD">Paso Doble</option>
            <option value="K">Kathak</option>
        </optgroup>
    </select>
    <input type="submit" value="Submit" />
</form>
```

This is just a simple form where all fields are mandatory and which collects information about the user—their name, e-mail ID, contact number, and which facility they wish to avail. As you must have noticed, we have inputs of types e-mail and number for the e-mail field and the contact number field respectively. The advantage of using the new input types on mobile websites is that when a user taps on a field of a particular type, the soft keyboard that is displayed will have keys according to the field. For example, if the user taps on the contact field, the soft keyboard that shows up will be a numeric only keypad. This makes the job easy for client-side validations as we eliminate the risk of a user entering wrong values to a great extent. We have added a `div` element after every input field to hold any error message.

We have the HTML ready, but to validate these fields, we need to have the validation JavaScript in place too. Add the following script within the `script` tag in body of the `index.php` file. Also make sure that you add it inside the `pagecontainershow` event at the very end after the existing code:

```
/*Form validation script begins*/
$("input").on('focus', function(){
    $(this).parent().next().text("");
    $(this).css('border-color', 'none');
});
$("input").on('blur', function(){
    var field_val = $(this).val();
    if(field_val === ""){
        $(this).parent().next().text("This field is mandatory.");
        $(this).css('border-color', 'red');
    }
});
$("input[type='text']").on('blur', function(){
    if($(this).val().match(/^[a-zA-Z]{1,20}$/) === null){
        $(this).parent().next().text("Name can only have alphabets and
maximum limit is 20 characters.");
        $(this).css('border-color', 'red');
    }
});
$("input[type='email']").on('blur', function(){
    if($(this).val().match(/^[a-zA-Z0-9._%+-]+@[a-zA-Z0-9.-]+\.
[a-zA-Z]{2,4}$/) === null){
        $(this).parent().next().text("Entered email is invalid.");
        $(this).css('border-color', 'red');
    }
});
$("input[type='number']").on('blur', function(){
    if($(this).val().match(/^[0-9]{,10}$/) === null){
```

```
        $(this).parent().next().text("Contact number can only be
numeric and should have 10 digits.");
        $(this).css('border-color', 'red');
    }
});
$("input[type='submit']").on('click', function(event){
    if($(".error").text() !== "" || $('.user input').val() === ""){
        alert("Make sure all fields are complete with no errors.");
        event.preventDefault();
        return false;
    }
});
/* Form validation script ends */
```

The form page with some field level validation errors appears, as shown in the following image, when you tap on the **Submit** button with some errors in the form:

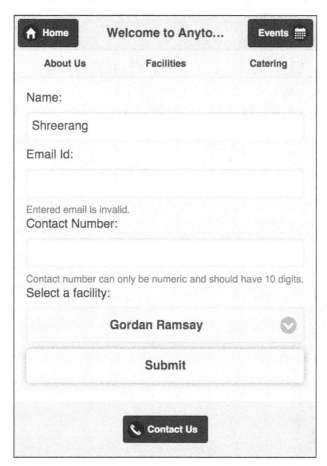

In validating the form, we are checking for the following validation rules:

- All fields are mandatory
- The name field can accept alphabets only
- The name field can accept a minimum of 1 character and a maximum of 20
- The e-mail field has to accept only the standard e-mail IDs
- The number field can accept only numeric values
- The length of the number field has to be 10
- If there are any errors on any of the fields, or if any of the fields are blank, then present a generic error and do not submit the form

In order to achieve all of these validations, we are making use of the blur event so that as soon as the user leaves the field, it is validated. Now that the form fields are validated, we can post the form values to a database table via PHP. We will take a look at this in the next section.

Inserting data into the database

To insert the form data into a database table, let's create a new table in our `Civic_Center` database. Let's call it `User_Queries`. This table will have the following five fields:

- An auto-increment ID field
- A field of type `varchar` to save the name of the user
- A field of type `varchar` to save the e-mail ID of the user
- A field of type `int` to save the contact number of the user
- A field of type `varchar` to save the value of the facility the user wishes to avail

Let's go over to `localhost/phpmyadmin` in your browser. Click on the **Civic_Center** database link in the left navigation. Click on the following **New** link it to add a new table. You will be presented with the following screen. Fill in the details as discussed earlier:

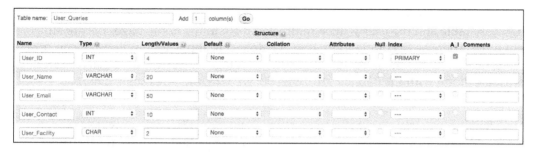

Click on the **Save** button at the bottom of the screen, and this will create your new table—User_Queries. We have now created a blank table that is ready to accept data values.

Create a new PHP page form.php, which will hold the PHP code to post the form values to the User_Queries database table. Add the following code to this file:

```php
<?php
    include 'db.php';
    // Check connection
    if (!$conn) {
        die("Connection failed: " . mysqli_connect_error());
    }

    $user_name = $_POST['name'];
    $user_email = $_POST['email'];
    $user_contact = $_POST['contact'];
    $user_facility = $_POST['facility'];

    $sql = "INSERT INTO User_Queries (User_Name, User_Email, User_
Contact, User_Facility) VALUES ('$user_name', '$user_email', '$user_
contact', '$user_facility')";

    if (mysqli_query($conn, $sql)) {
        header("Location: index.php");
    } else {
        echo "Error: " . $sql . "<br>" . mysqli_error($conn);
    }

    mysqli_close($conn);
?>
```

If the database connection is successful, the data from our form will be posted to the database table and the user will be redirected back to the home page. We are still not done. There are a couple more things before our application will be able to post the form data to the database table. On the `form` tag, we need to add the `action` attribute along with the `method` attribute. So, we will add `action="form.php"` `method="post"` to the `form` tag in the `index.php` file.

One final thing that we need to add to the `form` tag again is `data-ajax="false"`. In jQuery Mobile, form submissions are automatically handled using Ajax whenever possible, creating a smooth transition between the form and the result page. Unfortunately, this blocks the classic form handling via PHP. We can disable Ajax globally, but we need it for page handling and transitions. That's why adding `data-ajax="false"` on the `form` element is our best bet. After all these additions on the form element, it should look like the following:

```
<form class="user" novalidate action="form.php" method="post" data-ajax="false">
```

The application is now finally ready to post form data to the database. Fire up our application in the browser and head straight to the contact form from the home page. Fill in all the details correctly and click on the **Submit** button. If you get redirected to the **Home** page of our application, it means that the data was successfully added to the database table. You can verify this by checking the rows in the `User_Queries` table in **phpMyAdmin**:

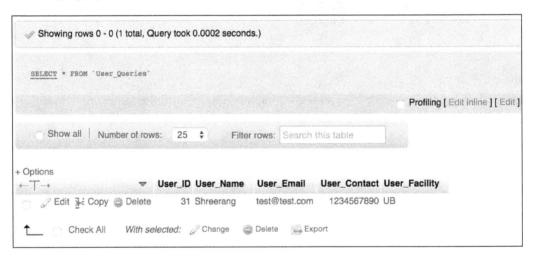

Summary

This was another exciting chapter, wasn't it? We started off by creating a MySQL database and created a table with some data using the `phpMyAdmin` interface. We saw how we can connect to this database using PHP and then we fetched data from the table to display it in the **Events** popup. We took this a step further and navigated the user from the events list to the details of the particular event that the user selected. Toward the end of the chapter, we created a new jQuery Mobile page that holds the contact form. We then created the JavaScript to do the client-side form validations and then saw how to post the form data to the MySQL database table using PHP.

In the next chapter, we will venture into the world of hybrid mobile applications using a framework called Cordova. We will create a simple but completely functional native mobile application using our jQuery Mobile code and some Cordova features. We will also discuss the important features of Cordova and how we can make use of them. Sounds interesting? Don't waste any time: quickly move on to the next chapter.

8
Finishing Touches

In the previous chapter, we discussed how we could connect our jQuery Mobile application with a backend database system to retrieve and show data dynamically. We saw how to set up a PHP-based system to fetch and save data to the database system. This made our system truly dynamic and real time.

Overview

In this chapter, we will take a look at how jQuery Mobile can be effectively used to create a hybrid mobile application. A hybrid mobile application is an installable mobile application developed using HTML, CSS, and JavaScript, and then wrapped inside a thin native container, Cordova; in our case, that provides access to the features of the native platform such as Android or iOS. We will look into the details of how to configure Cordova and look at some of the important features of Cordova in this chapter.

Creating custom icons

Before we move ahead and delve into the world of hybrid applications with the help of Cordova, we will take a look at an important aspect of jQuery Mobile — customizing the icons. jQuery Mobile comes with a predefined, out-of-the-box set of icons that are readily available and can be easily used in your jQuery Mobile application. However, on several occasions, we have to include custom icons that suit our application and are created by the design team in accordance with the design and style guide set for the application based on the customer needs. We will take a look at how we can incorporate such custom icons into our application and what the designers need to consider while creating an icon for any project that is being developed on the jQuery Mobile platform.

All the built-in icons that come along with the jQuery Mobile library are available in both SVG and PNG. So, basically, there is an SVG and a PNG image for every icon that is available by default. By default, the SVG icons are used, which render well on both retina and non-retina, SD and HD screen devices. The framework has an SVG support test in place, which adds a `ui-nosvg` class to the HTML element on platforms that don't support SVG. In such scenarios, the equivalent PNG images are displayed.

Ideally speaking, we need to create an SVG icon as well as a fallback PNG image for any custom icon that we need to use. However, in the real world, on a number of occasions, you might just receive a PNG or a JPEG image from the creative team, which is fine. The size of the default icons is 18 x 18 pixels. For your icon to be consistent with the other icons from the framework, it should be sized 18 x 18 pixels. However, this size will only support the non-retina or SD screen devices. We also need to support the retina and HD screen devices. For this, we need to create the icons sized 36 x 36 pixels.

To summarize, if we want to create a custom icon for use within a project based on jQuery Mobile platform, we need the following:

- An SVG and a PNG icon is needed. SVG icon is used by default by the framework and on non-SVG supporting devices; it falls back to displaying the PNG image
- The PNG icon should be sized at 36 x 36 pixels to support both retina and non-retina screen devices

Now, let's move on to how these custom icons can be incorporated within a jQuery Mobile application.

Icons are displayed as a background image of the `:after` pseudo elements. We need to target the pseudo element to set a custom icon. Let's take a look at the following code snippet to understand this better:

```
<button class="ui-btn ui-shadow ui-corner-all ui-btn-icon-left ui-icon-customicon">myicon</button>
```

Assuming that we are assigning a custom icon to a button, this is how the HTML should look like. We have added a class – `ui-icon-customicon` that will hold the custom icon as a background in the `:after` pseudo class. We will make use of the following CSS to achieve this:

```
.ui-icon-customicon:after {
    background-image: url("/images/custom_icon.svg");
}
```

```
/* Fallback */
.ui-nosvg .ui-icon-customicon:after {
    background-image: url("/images/custom_icon.png");
    background-size: 18px 18px;
}
```

Remember, SVG icons are used by default, and, if unavailable, the framework falls back to using PNG icons. If you have an SVG as well as a PNG image available for your icon, you will use the SVG image as a default, and to use the PNG image as a fallback, you need to add the fallback CSS with the `.ui-nosvg` class. In case you do not have the SVG image for your icon, just use the following CSS:

```
.ui-icon-customicon:after {
    background-image: url("/images/custom_icon.png");
    background-size: 18px 18px;
}
```

This is how you can create and implement a custom icon to spice up your jQuery Mobile application.

Introduction to Cordova

The documentation of Apache Cordova (`http://cordova.apache.org/docs/en/2.5.0/guide_getting-started_ios_index.md.html`) has the following to say:

> *Apache Cordova is a library used to create native mobile applications using Web technologies. The application is created using HTML, CSS and JavaScript and compiled for each specific platform using the platform native tools.*

This is how the Apache Cordova documentation defines the Cordova platform. From the definition, it is pretty evident that Cordova is some tool that makes use of our standard web development technologies, wraps it into something, and manages to make it available in form of a native mobile application. Any such native mobile application that is created using HTML, CSS, and JavaScript, wrapped using a native container, and made available as a native installable mobile application is termed as a hybrid mobile application. To understand what Cordova actually is and how it performs, we will have to step back a little and understand what and how hybrid applications work.

Hybrid mobile applications are same as any other applications available on your mobile device. These applications are installable and can be distributed to the public via the application stores. You can play games, capture pictures, make videos, engage with the world via social media, and pretty much do everything that you do using any of the other native mobile applications on your device. Like the websites on the Internet, hybrid applications are created using a combination of technologies such as HTML, CSS, and JavaScript. The only difference between the two is that instead of displaying this application in a browser, it is displayed within a mobile device's **Webview**. This allows the application to access the various device capabilities that are otherwise restricted to being used via web browser. The hybrid mobile applications can make use of device capabilities such as accelerometer, camera, GPS, and contacts among several others, and furthermore even make use of the native UI elements.

Now the question that arises is, how does this happen? How do the HTML, CSS and JavaScript-based applications gain access to the native features of a mobile device? The answer to these questions is Apache Cordova.

Apache Cordova is a free, open-source framework for building cross-platform native applications using HTML, CSS and JavaScript. It provides a simple way to create cross-platform mobile applications using a combination of web and native application technologies. The most important feature of Cordova is the native capabilities that it makes available which normally cannot be used from the mobile web browser. Cordova manages to provide a web application with native capabilities, by implementing a suite of APIs that extend the native capabilities of a device such as contacts list, accelerometer, and camera when it executes within a native container. Apache Cordova consists of the following components:

- Source code for the native application container for each of the supported mobile platforms: This container is the one that renders a Cordova-based web application on a mobile device

- A set of core JavaScript APIs that provide the web application running within this container and access to the native capabilities of the device platform

- A set of tools used to manage the process of creating and managing an application project, building native applications using the native software development kits—SDK—and, finally, testing the application using mobile device simulators and emulators

Apache Cordova does not translate the web application code written in HTML, CSS, and JavaScript into the native language for each of the supported device platforms. For example, Cordova does not translate the HTML code to Objective-C for iOS and Java for Android devices. In Cordova, the web application simply runs unmodified within a native application shell, with access to the underlying native platform features.

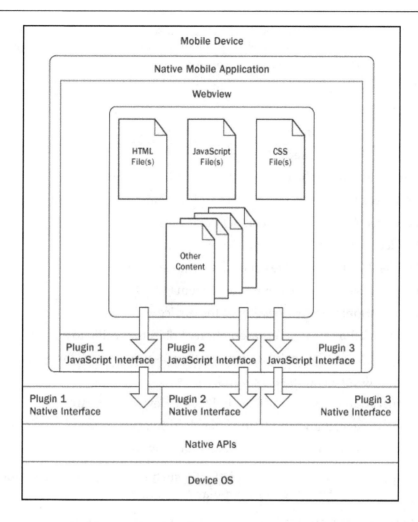

Cordova accommodates the need of web applications to access the native platform features by providing a suite of JavaScript APIs that are accessible to a web application running inside the Cordova native container **Webview**. Essentially, these APIs are implemented in two parts—a JavaScript library that exposes the native capabilities to the web application, and the corresponding native code running in the container that implements the native part of the API. This is implemented essentially as one JavaScript library, but has separate native implementations for each of the supported mobile device platforms.

Each of these Corodva APIs is made available as a separate plugin, which can be added or removed from your Cordova application using the Cordova **command-line interface (CLI)** or **plugin manager (plugman)**. Cordova currently provides the following APIs:

- **Accelerometer**: Tap into the device's motion sensor
- **Camera**: Capture images using the mobile camera
- **Capture**: Capture media files on the device
- **Compass**: Obtain the direction that the device is pointing to
- **Connection**: Check network state and cellular network information
- **Contacts**: Access device's contact database
- **Device**: Gather device specific information
- **Events**: Hook into native events using JavaScript
- **File**: Access the native device file system
- **Geolocation**: Capture location of the device
- **Globalization**: Enable representation of objects specific to a locale
- **InAppBrowser**: Launch URLs in another in-app browser instance
- **Media**: Record and play back audio files
- **Notifications**: Visual, audible, and tactile device notifications
- **Splashscreen**: Show and hide application splash screen
- **Storage**: Hook into device's native storage options

Cordova supports a variety of native platforms such as Tizen, FirefoxOS, Ubuntu, and many more apart from the most popular Android and iOS.

In a nutshell, Apache Cordova is a tool that provides an interface between the web technologies and the mobile device native technologies to help create an installable native mobile application on a variety of platforms. Now that we have a clear idea of what Cordova is and what its capabilities are, let's a take a look at how we can use it along with the jQuery Mobile library.

Configuring Cordova

In this section, we will take a look at how we set up our Mac system to build a Cordova-based application for Android and iOS. If you have a Mac machine, then you can develop and test your Cordova application for both Android and iOS. This is the reason why Mac is our choice for mobile application development. However, if you have a Windows box, then you can only build and test an Android application, and the installation process is pretty similar to the one on Mac. We will only take a look at the process you need to follow to set up your Mac machine for Cordova-based hybrid application development.

We will be installing XCode, iOS SDK, the **Command Line Tools**, and Android SDK required for the development process. Let's get started and discuss things in detail as and when the need be.

Step 1 – install XCode

Go to the Mac App Store and install XCode. It is free to download and essential for development on a Mac machine. The installation takes some time as the application size is huge. So, go, grab yourself a coffee while the installation process completes:

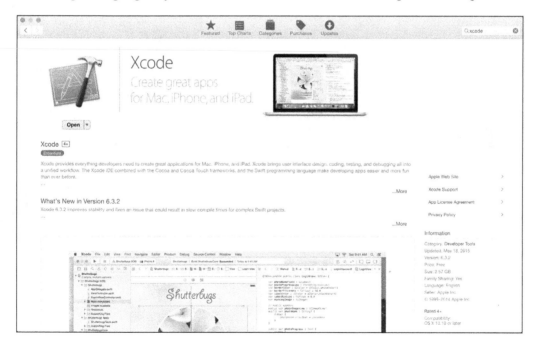

Step 2 – the iOS SDK

Once XCode is installed, you need to check if you have the latest iOS SDK available. To do this, go to the **XCode Preferences** and click on the **Downloads** tab, as shown in the following screenshot:

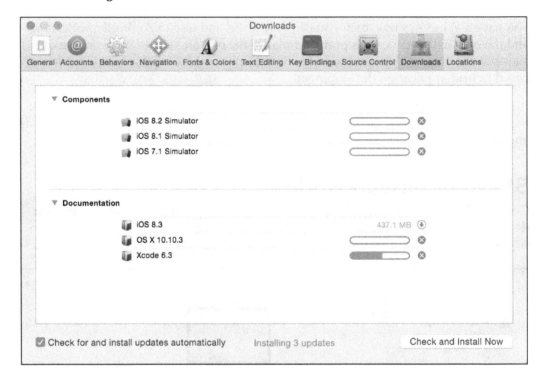

Step 3 – install Command Line Tools

To make sure that you have the Command Line Tools installed, open the **XCode preferences** and click on the **Locations** tab, as shown in the following screenshot:

Another way in which you can verify if the Command Line Tools are installed is to fire up your Terminal and type gcc. If available, you would see some text displayed on the Terminal. If the Command Line Tools are unavailable, the Terminal will give you a popup asking you to install the same.

Step 4 – install Android SDK

To develop Android applications, we need to install Java, the Ant build tool, and the Android SDK. To install Java, get the Java for OS X installer from Apple's download page:

Next, we need to install Ant. Ant is a build system for Java which is used by Cordova and Android SDK. Use Homebrew to install Ant:

```
ruby -e "$(curl -fsSL https://raw.githubusercontent.com/Homebrew/install/
master/install)"
brew install ant
```

Now we need to install the **Android Developer Tools (ADT)**. Download the ADT from `http://developer.android.com/sdk/index.html`. Android now provides us with the new Android Studio as the default, which replaces Eclipse with ADT. You can either download the Android Studio or you can still download just the SDK zip and install it on your machine to use with a different IDE. We will install the complete Android Studio:

Once you have installed the Android Studio, open it up and follow the steps in the Android Studio Installation Wizard. The installation is pretty quick and easy to follow.

Once the tools are installed you will need to create an environment variable for Android called `ANDROID_HOME`. You will create this using the following in your terminal:

```
export ANDROID_HOME=/Users/{your_username}/Library/Android/sdk
```

Then, add the following variable to your `path` as well as the path to the Android tools:

```
export PATH=$PATH:$ANDROID_HOME:/Users/{your_username}/Library/Android/
sdk/tools
```

One final step is to download the Android SDK version used by apache Cordova. To open the Android SDK Manager, go to the Terminal and simply type: `android`.

A window similar to the following screenshot should open up:

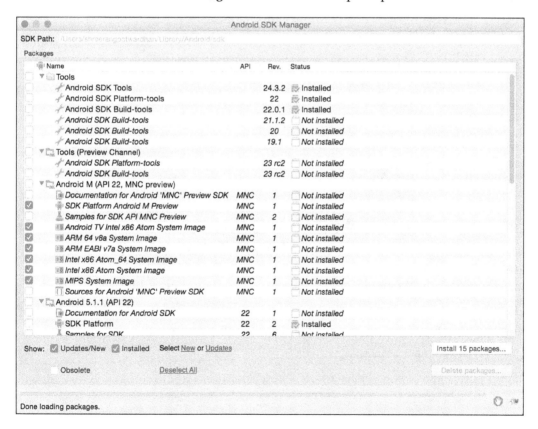

Select Android 4.2.2(API 19) SDK Platform and click on the **Install** package, as shown in the following screenshot:

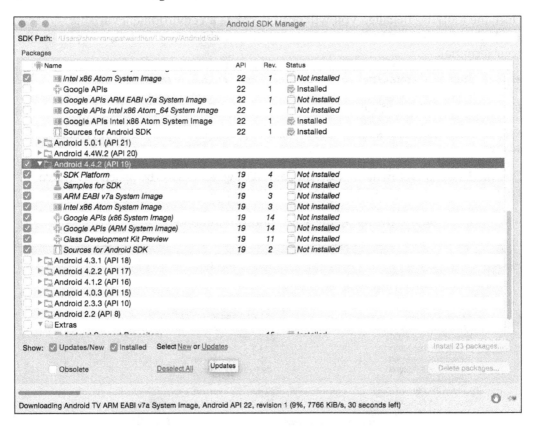

Step 5 – install Apache Cordova

Before installing Apache Cordova, you will need to install Node.js and Git on your Mac machine since Apache Cordova will be using Git to download some assets and you should be able to invoke node and npm on your command line to download Apache Cordova.

To install Apache Cordova, simply run the following command in your terminal:

```
sudo npm install -g cordova
```

Now, let us verify if Cordova has been installed correctly, and in the process of verifying this, we will also create our first Cordova application. This application shows nothing more than the default Cordova splash screen, but it is enough proof that our installation is successful and that we are ready to create Android and iOS applications using Cordova:

```
cordova create hello com.example.hello HelloWorld
```

Here's the description of the preceding command:

- `hello` is the name of the directory you want to create your project in.
- `com.example.hello` provides your project with a reverse domain-style identifier.
- `HelloWorld` provides the application with a display title. This can be edited at any time in the `config.xml` file.

Now, the next step is to add platforms for which we wish to build our application:

```
cd hello
cordova platform add ios
cordova platform add android
```

Once you have added the platforms, we can now build our application. The application will be built for all the platforms you have added:

```
cordova build
```

After the build is successful, we would want to see the application in action. To do this, we need to fire up the emulator for Android and iOS. Before running the iOS emulator, you will need to install `ios-sim`. To install `ios-sim`, run the following:

```
$ git clone git://github.com/phonegap/ios-sim.git
cd ios-sim
$ rake install prefix=/usr/local/
```

We can now run the iOS emulator:

```
cordova emulate ios
```

And there you go:

To run the emulator for Android:

```
cordova emulate android
```

On firing this command, the Android emulator will come up and should be as shown in the following image:

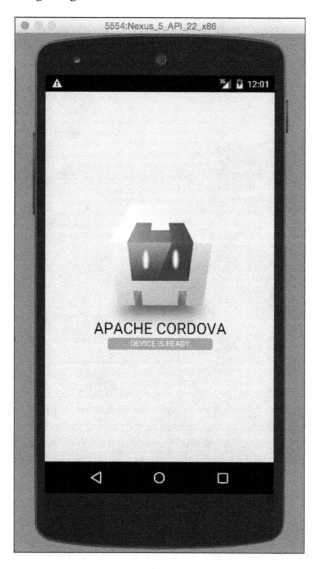

Now that we have Cordova installed and firing up fine, let's take a breather, might as well fill up our coffee mugs and get set to create a really simple, yet functional hybrid application using jQuery Mobile. We will reuse part of the code that we already have in place and then modify the Hello World application that we just created. Let's take a look at the folder structure of this application and then we can discuss about which files we need to modify:

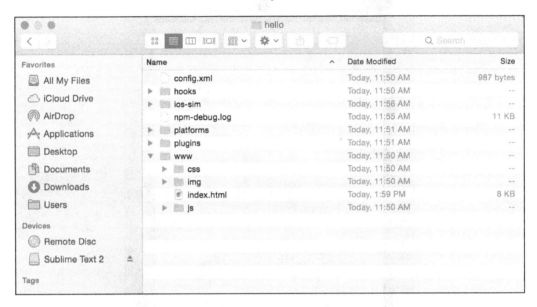

Like in our web application so far, we will be modifying the index.html file in the www folder, any custom JavaScript goes into the js folder, and all custom CSS goes into the css folder. For creating the hybrid application, we will use the HTML for the facilities page, which means we will be seeing a mobile native application that shows the facilities page from our Civic Center application.

Fire up the index.html file from our Civic Center application and the index. html file from the Hello World Cordova application. You will notice that there are several defaults in the index.html file from the Cordova application. We are going to replace it all, but we encourage you to go through the details of this file. From the index.html file of our Civic Center application, copy only the facilities page HTML and place it inside of div with class application. Include the jQuery Mobile CSS file reference in the header as part of the head tag. Include the jQuery and jQuery Mobile JavaScript file references at the very end of the document just before the close of the body tag.

Your header section, the `head` tag, should now look like the following:

```
<head>
        <meta name="format-detection" content="telephone=no">
        <meta name="msapplication-tap-highlight" content="no">
        <meta name="viewport" content="width=device-width; initial-
scale=1.0; maximum-scale=1.0; minimum-scale=1.0; user-scalable=no;" />
        <link rel="stylesheet" href="http://code.jQuery.com/
mobile/1.4.5/jQuery.mobile-1.4.5.min.css" />
        <title>Hello World</title>
    </head>
```

The body of this file, where the main crux of our application lies, should include the following:

```
<div class="app">
        <!-- Facilities page beigns-->
        <div data-role="page" id="facilities">
            <div data-role="popup" id="eventregister3" data-
overlay-theme="b" data-theme="b" data-dismissible="false">
                <div data-role="header" data-theme="a">
                    <h1>Current Events</h1>
                </div>
                <div role="main" class="ui-content">
                    <h3 class="ui-title">Sorry, currently there
are no scheduled events.</h3>
                    <a href="#" class="ui-btn ui-corner-all ui-
shadow ui-btn-inline ui-btn-b" data-rel="back">Close</a>
                </div>
            </div>

            <div data-role="header">
                <a href="#home" class="ui-btn-left ui-btn ui-
btn-inline ui-mini ui-corner-all ui-btn-b ui-btn-icon-left ui-icon-
home">Home</a>
                <h1>Welcome to Anytown Civic Center!</h1>
                <a href="#eventregister3" data-rel="popup" data-
position-to="window" data-transition="pop" class="ui-btn-right ui-
btn-b ui-btn ui-btn-inline ui-mini ui-corner-all ui-btn-icon-right ui-
icon-calendar">Events</a>
                <div data-role="navbar">
                    <ul>
                        <li><a href="#about">About Us</a></li>
                        <li><a href="#facilities" data-
ajax="false">Facilities</a></li>
```

```
                              <li><a href="#catering">Catering</a></li>
                        </ul>
                  </div><!-- /navbar -->
            </div><!-- /header -->


            <div data-role="panel" id="contactpanel" data-
display="push" data-dismissible="true" data-theme="a">
                  <div>
                        <p>Contact Us!</p>
                        <a href="tel:555-555-5555">(555)555-5555</a>
                        <p>1234 First Avenue</p>
                        <p>Anytown, Anystate 12345</p>
                        <a href="mailto:contact@anytownciviccenter.
com">contact@anytownciviccenter.com</a>
                  </div>
            </div><!-- /panel -->


            <div class="ui-content" role="main">
                  <p>At Anytown Civic Center we have the following
facilities at your disposal:<br />
                        <div data-role="collapsibleset">
                              <div data-role="collapsible">
                                    <h3>Banquet Halls</h3>
                                    <p>
                                          <span>
                                                We have 3 huge banquet halls
named after 3 most celebrated Chef's from across the world.
                                          </span>
                                          <ul data-role="listview" data-
inset="true">
                                                <li><a href="#">Gordon
Ramsay</a></li>
                                                <li><a href="#">Anthony
Bourdain</a></li>
                                                <li><a href="#">Sanjeev
Kapoor</a></li>
                                          </ul>
                                    </p>

                              </div>
                              <div data-role="collapsible">
                                    <h3>Sports Arena</h3>
                                    <p>
                                          <span>
```

We have 3 huge sport arenas named after 3 most celebrated sport personalities from across the world.

```
                                            </span>
                            <ul data-role="listview" data-
inset="true">
                                    <li><a href="#">Sachin
Tendulkar</a></li>
                                    <li><a href="#">Roger
Federer</a></li>
                                    <li><a href="#">Usain Bolt</
a></li>
                            </ul>
                    </p>
            </div>
            <div data-role="collapsible">
                    <h3>Conference Rooms</h3>
                    <p>
                            <span>
                                    We have 3 huge conference
rooms named after 3 largest technology companies.
                            </span>
                            <ul data-role="listview" data-
inset="true">
                                    <li><a href="#">Google</a></
li>
                                    <li><a href="#">Twitter</a></
li>
                                    <li><a href="#">Facebook</a></
li>
                            </ul>
                    </p>
            </div>
            <div data-role="collapsible">
                    <h3>Ballrooms</h3>
                    <p>
                            <span>
                                    We have 3 huge ball rooms
named after 3 different dance styles from across the world.
                            </span>
                            <ul data-role="listview" data-
inset="true">
                                    <li><a href="#">Ballet</a></
li>
                                    <li><a href="#">Kathak</a></
li>
```

```
                                        <li><a href="#">Paso Doble</
    a></li>
                                </ul>
                            </p>
                        </div>
                    </div>
                    Contact us for pricing and availability.
                </p>
            </div><!-- /content -->

            <div data-role="footer">
                <h2>
                    <a href="#contactpanel" data-rel="panel"
    class="ui-btn ui-btn-inline ui-mini ui-corner-all ui-btn-b ui-btn-
    icon-left ui-icon-phone">Contact Us</a>
                </h2>
            </div>
        </div><!-- /page -->
        <!-- Facilities page ends -->
    </div>
    <script type="text/javascript" src="cordova.js"></script>
    <script src="http://code.jQuery.com/jQuery-1.10.1.min.js"></
    script>
    <script src="http://code.jQuery.com/mobile/1.4.5/jQuery.
    mobile-1.4.5.min.js"></script>
```

You must have noticed that we have a `cordova.js` file at the end of our `index.html` file here. This JavaScript file basically helps our web application to communicate with the native platform technologies to help us create a native application. We now have the code ready and it is ready to be compiled and executed. Follow the commands below to fire up the android and iOS emulators:

```
cd
cd hello
cordova build
cordova emulate ios
cordova emulate android
```

The output is extremely satisfying, as you see the **Facilities** page fire up in a native application. The output on the emulators will be as shown in the following images:

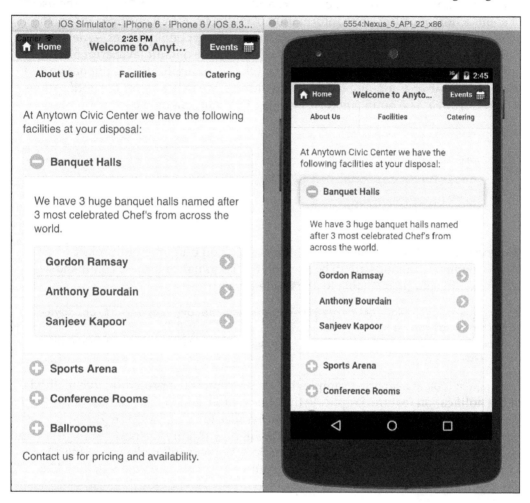

This is just a demonstration of how jQuery Mobile and Cordova can be used together to create native applications. The possibilities here are endless and sky is the limit for experimentation here. We encourage you to explore this amazing framework further and create awesome applications using these two frameworks and even try to convert our entire Civic Center application to a hybrid application. We will now look at a few prominent features of this framework in the further sections. If you are really interested in the Cordova framework and want to explore it further, we encourage you to look into one of the several titles available from Packt Publishing on this topic.

Notifications

In a mobile application, on numerous occasions, we would want to notify the user of some action or collect some information from the user. In web applications, to achieve this functionality we make use of the JavaScript popup boxes — Alert Box, Confirm Box, and Prompt Box. If we make use of these default JavaScript alerts in our native application, it gives away the fact that your application is not native and gives the end-user a non-native feel. To avoid this, we make use of the notifications API in Cordova. The notification API has the following methods:

- `notification.alert`
- `notification.confirm`
- `notification.prompt`
- `notification.beep`
- `notification.vibrate`

To access this feature, we need to add the notification plugin to our project. The following commands are to be fired on the Terminal, and these commands will make this plugin available to all targeted platforms:

```
$ cordova plugin add https://git-wip-us.apache.org/repos/asf/cordova-plugin-vibration.git
$ cordova plugin add https://git-wip-us.apache.org/repos/asf/cordova-plugin-dialogs.git
```

Let's take a look at a simple example that demonstrates the preceding five methods of the notification feature. Duplicate the `index.html` and `index.js` files of the `Hello World` Cordova project and rename the earlier files to `jQueryMobile_demo.html` and `jQueryMobile_demo.js`, respectively. We will now modify the new `index.html` and `index.js` files in the same `Hello Work` project. We will still make use of the jQuery Mobile framework to style our UI, and so our new `index.html` file will look like the following:

```html
<html>
    <head>
        <title>Notification Example</title>
        <meta name="viewport" content="width=device-width; initial-scale=1.0; maximum-scale=1.0; minimum-scale=1.0; user-scalable=no;" />
        <link rel="stylesheet" href="http://code.jQuery.com/mobile/1.4.5/jQuery.mobile-1.4.5.min.css" />
    </head>
  <body>
    <p>
        <ul data-role="listview" data-inset="true">
```

```
        <li><a href="#" onclick="showAlert(); return
false;">Native Alert Box</a></li>
            <li><a href="#" onclick="showConfirm(); return
false;">Native Confirm Box</a></li>
            <li><a href="#" onclick="showPrompt(); return
false;">Native Prompt Box</a></li>
        </ul>
    </p>
    <script type="text/javascript" charset="utf-8" src="cordova.js"></
script>
    <script src="js/index.js"></script>
    <script src="http://code.jQuery.com/jQuery-1.10.1.min.js"></
script>
    <script src="http://code.jQuery.com/mobile/1.4.5/jQuery.mobile-
1.4.5.min.js"></script>
  </body>
</html>
```

As you must have noticed, we have included the `index.js` file. This file holds our custom JavaScript code, which is required for the execution of this demo. The `javascript` file contains simple, self-explanatory functions, and the file contents are as follows:

```
document.addEventListener("deviceready", onDeviceReady, false);

/* Function to display alert message*/
function showAlert() {
    navigator.notification.alert(
        'You are the winner!',  // message
        null,                   // callback
        'Game Over',            // title
        'Done'                  // buttonName
    );
    playBeep();
    vibrate();
}

/* Function to play a beep sound*/
function playBeep() {
    navigator.notification.beep(1);
}

/* Function to make the device vibrate*/
function vibrate() {
    navigator.notification.vibrate(2000);
}
```

```
/* Function to display a confirm dialog box*/
function showConfirm() {
    navigator.notification.confirm(
        'You are the winner!',  // message
        onConfirm,              // callback to invoke with index of
button pressed
        'Game Over',            // title
        'Restart,Exit'          // buttonLabels
    );
}


/* The confirm dialog callback function*/
function onConfirm(buttonIndex) {
    var butttonText = '';
    if (buttonIndex == 1){
        butttonText = 'Restart';
    }
    else{
        butttonText = 'Exit'
    }
    alert('You selected the ' + butttonText + ' button.');
}


/* Function to display a prompt box*/
function showPrompt() {
    navigator.notification.prompt(
        'Please enter your name',  // message
        onPrompt,                  // callback to invoke
        'Registration',            // title
        ['Ok', 'Exit'],            // buttonLabels
        'Jane Doe'                 // defaultText
    );
}


/* The prompt dialog callback funtion*/
function onPrompt(results) {
    alert('You entered ' + results.input1);
}
```

To compile and run this Cordova project, fire up your terminal and implement the following commands:

```
cordova build
cordova emulate ios
cordova emulate android
```

The output of this simple application as seen in the iOS emulator is as shown in the following screenshot:

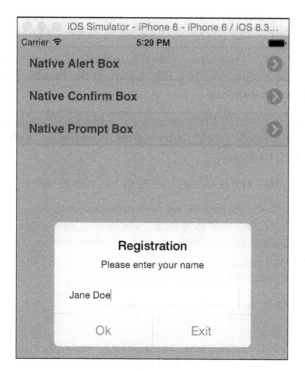

Geolocation

Geolocation is one the most important and a very popular feature of a smart phone device. You must have used several geolocation-based apps such as Google Maps Navigation, Foursquare, Facebook check-in, or Yelp. Geolocation is the basis of all these applications. Geolocation provides the information about the device's location such as its latitude and longitude. The location information is collected via the usual means such as the **Global Positioning System** (**GPS**) module available in the device and the location inferred from network signals such as IP addresses, RFID, Wi-Fi and Bluetooth MAC addresses, and GSM/CDMS cell IDs. There is no guarantee that the API returns the exact location of the device, but it depends on various factors.

The geolocation API has the following methods:

- `geolocation.getCurrentPosition`
- `geolocation.watchPosition`
- `geolocation.clearWatch`

To access the geolocation feature, we need to add this plugin to our project using the following command on the Terminal:

```
$ cordova plugin add https://git-wip-us.apache.org/repos/asf/cordova-plugin-geolocation.git
```

We will not build a full blown example for this feature. However, we will take a look at the crux of this feature. We will take a look at the function we would need to capture the location of a device.

We will wait for the device API libraries to load, and to do that, add a event listener to check if the device is ready. As in the previous example, we will make use of the the following function to do so:

```
document.addEventListener("deviceready", onDeviceReady, false);
```

Within this `onDeviceReady` function, we will call geolocation's `getCurrentPosition` method. We will include two callback functions in this: one that executes when it receives the exact latitude and longitude values of the device and the other one executes in case the API is unable to detect the location of the device; in that case it receives a `PositionError` object.

The success function would look something like:

```
function onSuccess(){
    var lat = position.coords.latitude;
    var long = position.coords.longitude;
    var alt = position.coords.altitude;
    var accuracy = position.coords.accuracy;
    var altAccuracy = position.coords.altitudeAccuracy;
    var head = position.coords.heading;
    var speed = position.coords.speed;
    var timestamp = position.coords.timestamp;
}
```

We receive these 8 different properties of the position object when the API returns successfully. Depending on whether your application is a mapping application that needs to display the current location of the user on the map or it is an activity tracking application that needs the altitude as well apart from the current location, you can include these properties within your application.

The watchPosition method of geolocation is used to watch for changes in the device's current location. This method proves extremely useful when creating a navigation-based or activity tracking application where a change in location of the user's device is to be tracked. The clearWatch method on the other hand is to be used when you want to stop looking for the device's location.

Geolocation is the next buzzword, and several creative applications have been created based on this and will be created in the future as well. However, always remember, with great power comes great responsibility. When your application makes use of geolocation to track a device's location, you should make sure you are not infringing the user's privacy. Geolocation data is generally considered to be sensitive because it can reveal the whereabouts of a particular user, and, if saved, the history would reveal the user's travel details. You should always make sure that you have included all such things related to location tracking in the privacy policies of your application and have intimated the user that the application makes use of location tracking services. Hope you are already thinking of different ways in which you would use the Geolocation API from Cordova and utilize it in a meaningful way and create some cool UI for your application using Google Maps JavaScript API and jQuery Mobile.

The next and the final feature of Cordova that we will be looking at as part of this chapter is *Offline storage*. We will take a look at this in detail in the next section.

Offline storage

Cordova being a client side framework has access to the exact same storage options as any other HTML5-based web application. However, based on the device and the underlying platform that you are using, the support might vary. Bearing this fact in mind, following are the various options that are available to you to select and make use of in your mobile application:

- **Local storage**: You can save private primitive data in key-value pairs and is generally the best option to save data locally and offline.

- **WebSQL**: The WebSQL API is available to underlying Webview. The Web SQL Database specification offers database tables that can be accessed via SQL queries. Not all platforms support WebSQL.

- **IndexedDB**: IndexedDB such as WebSQL is available in the underlying Webview. IndexedDB offers more features than local storage but fewer than WebSQL. IndexedDB too is not supported by all the popular platforms.

- **Plugin-based options**: In addition to the WebSQL and IndexedDB storage APIs, the file API allows you to cache data on the local file system.

We will now take a look at two of these data storage options that are supported by Cordova and can also be used to store data offline.

Local storage

Local storage is available to all modern browsers and is pretty simple to use. All of the data is stored in the global `localStorage` attribute, which is an associative array, and so each item is referenced using a key. Data is stored to local storage using the `setItem()` method, which accepts a key and corresponding value:

```
window.localStorage.setItem('user_id', 'test123');
```

This data can be saved to `localStorage` even when your mobile application is offline. You can detect the presence of a network using the Connection API of Cordova, and once the device is connected to a data network or Wi-Fi, you can easily retrieve the information from the `localStorage` and post it to the backend database using the web services that you might have in place. Using `localStorage` you can save up to 5 MB of data within your mobile application. To retrieve the data from `localStorage`, we need to make use of the `getItem` method rather than the `setItem` method:

```
window.localStorage.getItem('user_id');
```

WebSQL database

WebSQL databases are well supported by the latest browsers. If you have prior experience working with databases and SQL queries, you will find this implementation quite familiar. The following code snippet is a simple implementation of the WebSQL SQL queries. The first query creates a user table and inserts a few rows into it:

```
var db = openDatabase('testDb', '1.0', 'User DB', 2*1024*1024),
    msg = '',
    status = document.getElementById('db-data');

db.transaction(function (tx) {
    tx.executeSql('CREATE TABLE IF NOT EXISTS USER (id unique,
user)');
    tx.executeSql('INSERT INTO USER (id, log) VALUES (1, "Test User
1")');
    tx.executeSql('INSERT INTO LOGS (id, log) VALUES (2, "Test User
2")');
    msg = '<p>Log message created and row inserted.</p>';
    status.innerHTML =  msg;
});
```

In the transaction that follows, we fetch the contents of the user table and display them in a webpage:

```
db.transaction(function (tx) {
    tx.executeSql('SELECT * FROM USER', [], function (tx, results) {
        var len = results.rows.length, i;
            msg = "<p>Found rows: " + len + "</p>";
        status.innerHTML +=  msg;
        for (i = 0; i < len; i++) {
            msg = "<p><b>" + results.rows.item(i).user + "</b></p>";
            status.innerHTML +=  msg;
        }
    }, null);
});
```

WebSQL database too, like `localStorage`, is available when the mobile application is offline. The application can save data to the database even when there is no network connectivity and, then, once the network connectivity is available, can push this data to the remote, more permanent database.

With this, we come to the end of this chapter and conclude our discussion related to Apache Cordova and how jQuery Mobile can be used with it to create beautiful looking native mobile applications.

Summary

We started off this chapter by looking at how we can create custom icons for jQuery Mobile, what factors we need to consider while creating these icons, and how to implement them in our jQuery Mobile application. We then shifted our focus to hybrid mobile applications and Apache Cordova for the rest of the chapter. We discussed what hybrid applications are, and we understood the need for these hybrid applications. Apache Cordova is one such framework that provides the interface between web technologies and the native platform technologies. We looked indepth how to set up our Mac system for mobile application development, installing Cordova, iOS and Android SDKs, and the Command Line Tools. We even created a simple Cordova and jQuery Mobile-based application using our Civic Center application code. Toward the end, we discussed a few, most important features of Cordova – notifications, Geolocation, and Offline storage.

In the next chapter, we will take a look at some more fun ways in which we can use jQuery Mobile. We will make use of jQuery Mobile in a Node.js-based application along with RequireJs and Backbone.js. This will basically give us an idea of how the jQuery Mobile framework works along with the modern MVC frameworks. Toward the end of the next chapter, we will also create a WordPress theme using jQuery Mobile. Sounds interesting, right? Take a coffee break and come back to explore the fun as you turn the page. We will wait for you...

9
The Next Level

We have finished the main project of this book and then revved it up by turning a small part of it into a native app with Apache Cordova, but the fun is not over yet! In this chapter, we will create a few advanced standalone projects to show you various other ways of using jQuery Mobile.

Overview

The goal of this chapter is to show you some additional ways in which you can use jQuery Mobile and master it. Thus, in this chapter, we will look at the following topics:

- Working with Node.js
- Working with RequireJS and Backbone.js
- Building a WordPress Mobile Theme

Each of these projects will give you some additional real-world practice with jQuery Mobile and show you how it can be integrated into almost anything on the Internet.

 These projects will be small and are only meant to show you the basics of integrating jQuery Mobile with other technologies. There will be a small introduction, but no grand details on how to set up a Node.js server, for example, or build a responsive e-commerce theme for WordPress. We will, however, try to recommend a book on every topic so that you can learn more as you want.

Working with Node.js

Node.js is an open source, cross-platform runtime environment for server-side and networking applications. This is its official description, and basically, it allows for server-side JavaScript. Due to its speed, it is used primarily for real-time data-intensive applications. Using Node.js, a developer can write a complete application in JavaScript without having to rely on using a server-side language, such as PHP, ASP.NET, and the like.

Getting jQuery Mobile to play nicely with Node.js applications takes a bit of work, but it is completely doable.

 Again, we are not going into great detail on how to get Node.js up and running on your server or local development machine. We are going to assume you already have Node.js running and are familiar with developing/running applications that use it. Also for Node.js development, we are huge fans of the node tools offered in Visual Studio 2013, so we will be using them in this section. You can download the free Visual Studio 2013 Community Edition and the Node.js tools for free and follow along if you like. Otherwise, use the IDE you are comfortable with.

To make things easy for this project, we will be using Express and Jade packages, so be sure to install them before we get started. Here is the directory structure of our project:

Getting started

Create your Node.js project (don't forget to add the express and jade modules). Here are the contents of our package.json file:

```
{
    "name": "NodejsWebApp1",
    "version": "0.0.0",
    "description": "NodejsWebApp1",
    "main": "server.js",
    "author": {
        "name": "chip",
        "email": ""
    },
    "dependencies": {
        "express": "*",
        "jade"  : "*"
    }
}
```

This is a typical package.json file. We specify some information about our project, such as the name, version, and author information, and then we declare our dependencies.

Creating a starting JavaScript file

Add the following code to the main file (server.js in this case):

```
var express = require('express');
var routes = require('./routes');
var http = require('http');
var port = process.env.port || 1337;
var jqm = express();

jqm.set('port', process.env.PORT || 1337);
jqm.set('views', __dirname + '/views');
jqm.set('view engine', 'jade');

jqm.get('/', routes.index);
jqm.get('/home', routes.index);

http.createServer(jqm).listen(jqm.get('port'), function () {
    console.log('URL: http://localhost:' + jqm.get('port'));
});
```

This code holds several configuration variables we need for our project. Here, we specify the port we want our project to run on, we require `express`, we set our view engine to `jade`, and we create our routes.

Creating our jade views

Now we come to the fun part! We will create two `jade` views. We will create one base layout template file and then one that will expand on that view. This way you can grow your project, and if you have to make drastic layout changes, you can change one file instead of changing multiple ones.

The first Jade file we will create is the main layout view. So add a new file to the `layout.jade` project (be sure to place it in the `views` folder) and add the following code to it:

```
doctype html
html
  head
    title= title

    meta(http-equiv="Content-Type", content="text/html;
charset=utf-8")
    meta(name="apple-mobile-web-app-capable", content="yes")
    meta(name="viewport", content="width=device-width, initial-
scale=1.0")

    link(rel="stylesheet" href="http://code.jquery.com/mobile/1.4.5/
jquery.mobile-1.4.5.min.css")

    script(src="http://code.jquery.com/jquery-1.9.1.min.js")
    script(src="http://code.jquery.com/mobile/1.4.5/jquery.mobile-
1.4.5.min.js")
  body
    div(data-role="page")
    div(data-role="header", data-position="fixed")
    h2= title
    a(href="/home", data-icon="home", data-ajax="false", class="ui-
btn-left") Home
    div(role="main", class="ui-content")
    block content
    div(data-role="footer", data-position="fixed")
    h3 &copy; 2015 Your Copyright
```

You should, for the most part, recognize the majority of this code; but since it has been presented slightly differently in Jade, let's discuss it in detail.

Jade is a template engine for Node.js. It allows for shorthand HTML tagging—this is why some of this code may look different from the traditional HTML code you were raised with.

The following code snippet is your standard HTML opening tag information, just presented without brackets. Since we will specify the page title through our routes, we use the title variable, which we will change on the fly:

```
doctype html
html
     head
          title= title
```

In the following code section, we set some required `meta` tags, including a couple to make the app more user friendly on a mobile device. The apple-mobile-web-app-capable tag will allow this app to function well on iOS devices in full-screen mode:

```
meta(http-equiv="Content-Type", content="text/html; charset=utf-8")
meta(name="apple-mobile-web-app-capable", content="yes")
meta(name="viewport", content="width=device-width, initial-scale=1.0")
```

In the following code, we make our necessary jQuery Mobile framework calls to pull in the framework itself and the CSS files:

```
link(rel="stylesheet" href="http://code.jquery.com/mobile/1.4.5/
jquery.mobile-1.4.5.min.css")

     script(src="http://code.jquery.com/jquery-1.9.1.min.js")
     script(src="http://code.jquery.com/mobile/1.4.5/jquery.mobile-
1.4.5.min.js")
```

The following is the main piece of our template. Here, we create our jQuery Mobile page, specifying the header and footer. Now, there is one piece that may seem off to you if you are not familiar with Jade, and that is `block content`. Much like our title declaration, this too is a placeholder. We will override that with actual content blocks in other template files, which we will create in a bit:

```
body
   div(data-role="page")
     div(data-role="header", data-position="fixed")
       h2= title
       a(href="/home", data-icon="home", data-ajax="false", class="ui-
btn-left") Home
     div(data-role="content")
       block content
     div(data-role="footer", data-position="fixed")
       h3 &copy; 2015 Your Copyright
```

Add another file to your project named `index.jade` (again, be sure to place it in the `views` folder). For all intents and purposes, this will be our `index.html` file. Once you have created the file, add to it the following code:

```
extends layout
block content
  h2= title
  p.
    Hello World!
```

This is not a lot of code, but with the way Jade works, our subsequent pages don't need a lot of code. The following line pulls in our `layout.jade` file. Now whenever you make changes to `layout.jade`, they will be reflected in every file that has this line:

```
extends layout
```

What happens here is that the preceding code block gets placed or injected at all places where we had the block content line in `layout.jade`. Notice that to display basic text, you must have `p.`; think of it as an opening `<p>` tag in HTML:

```
block content
  h2= title
  p.
    Hello World!
```

One thing you might have noticed in our code with Jade is, that there are no closing tags like we have in HTML. That is because they are not needed in Jade. Jade is whitespace-sensitive, so it automatically closes the tags for us.

 Interested in learning more about Jade? It truly is a great template engine and we highly recommend you learn more about it. We recommend checking out the book *Web Development with Jade, Sean Lang, Packt Publishing*. You can find it at `https://www.packtpub.com/web-development/web-development-jade`.

Creating our routes

Now, we have our Node.js application specified and we have created our views; now to create the routes that will hold everything together, so to speak.

Create a new JavaScript file named `index.js` in the routes folder and enter the following code in it:

```
exports.index = function (req, res) {
    res.render('index', { title: 'jQuery Mobile and Node'});
};
```

In this file, we create our route to render our index template and we pass it the title of jQuery Mobile and Node.js so that it will show up on the page. Since we are not doing anything overly complicated (interfacing with MongoDB or such), this is pretty much it for this file. If you were to connect to MongoDB, you would place all the required code here.

Now we have finished our Node.js application; so go ahead and execute it. You should see a message as in the following screenshot:

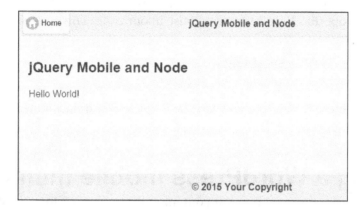

Congratulations! You just got Node.js and jQuery Mobile to play nicely together! Node.js is a wonderful alternative to using server-side languages, such as PHP or .NET. If you are comfortable with JavaScript, you should check it out.

> Packt has several titles on Node.js development. To get started, we recommend *Node Web Development - Second Edition, David Herron, Packt Publishing*, found at `https://www.packtpub.com/web-development/node-web-development-second-edition`.

Working with RequireJS and Backbone.js

RequireJS and Backbone.js are two extremely popular JavaScript libraries. RequireJS is known as an asynchronous module definition script loader library that can improve page-load time as well as assist with dependency management. Backbone.js is a popular MVC framework. Both of these work great with jQuery Mobile, provided you get them to play together nicely.

To start getting into development with all three libraries, you have to turn off a couple of jQuery Mobile's properties. These two properties are:

- `$.mobile.linkBindingEnabled`: This property allows jQuery Mobile to bind clicks to your anchor tags on the page.

- `$.mobile.hashListeningEnabled`: This property enables jQuery Mobile to listen and handle changes to `location.hash`.

In order for Backbone.js or RequireJS to function properly with jQuery Mobile, both of these properties have to be set to false in our code. This code simply looks like the following:

```
$.mobile.linkBindingEnabled = false;
$.mobile.hashListeningEnabled = false;
```

Once you have these turned off, you can use RequireJS and Backbone.js as you normally would.

Building a WordPress mobile theme

WordPress is one of the most popular **content management systems (CMS)** on the Web. Developed in PHP by Matt Mullenweg and Mike Little back in 2003, it started out as a blogging platform, but has since evolved into a very powerful web platform.

One of the alluring factors of WordPress is its powerful plugin and theme systems. In this section, we will create a mobile theme for WordPress powered by jQuery Mobile.

Getting ready

The first thing to do is make sure you have a WordPress installation ready to access. We personally recommend installing it locally on the development server you have been using throughout this book. You may download the package from http://www.wordpress.org. Installation is simple. Just unzip the file into your `htdocs` directory, bring up the directory in your browser, and follow the onscreen prompts. If you need additional assistance beyond this crash course, please refer to the documentation on the WordPress website.

Now that you have a working copy of WordPress, let's look at what we need to create the theme. Each WordPress theme must have `index.php` and `style.css` files at least. For the purpose of this project, we are going to have the following file structure:

- `index.php`
- `header.php`

- `footer.php`
- `functions.php`
- `style.css`

You can create these files wherever you like. However, we recommend creating them in `wp-content/themes/jqmwptheme`, since this is where your theme will be once we are finished.

Creating the style sheet

The `style.css` file is extremely important, as it contains information about our theme that will be displayed in the administration section of the WordPress dashboard. In Aptana Studio or your IDE of choice, go ahead and create a new file named `style.css`, and add the following code to it:

```
/*
Theme Name: My Awesome jQuery Mobile Theme
Theme URI: http://localhost/myawesomejqmtheme
Author: Chip Lambert (or your name here)
Author URI: http://programmerchip.com
Description: This theme is an extremely basic Wordpress theme showing
the jQuery Mobile framework. Feel free to use this as a basis for your
own awesome theme!
Version: 1.0
License: GNU General Public License v2 or later
License URI: http://www.gnu.org/licenses/gpl-2.0.html
Tags: chip, lambert, awesome, jquery mobile, mobile, world domination,
wordpress

This theme, like WordPress, is licensed under the GPL.
Use it to make something cool, have fun, and share what you've learned
with others.
*/
```

Save the file. This is all that we will have in this file, as our actual CSS will come from jQuery Mobile.

In this file, we create the essential information needed by the WordPress application for the theme to be usable. If this information is not present, WordPress will not recognize this as a valid theme. Also, this information must be unique, meaning no other theme can have this exact same information.

Creating the header file

Now let's create the `header.php` file and put the following code in it:

```
<!DOCTYPE html>
<html <?php language_attributes(); ?>>
  <head>
    <meta charset="<?php bloginfo( 'charset' ); ?>" />
    <title><?php wp_title(); ?></title>
    <link rel="profile" href="http://gmpg.org/xfn/11" />
    <link rel="pingback" href="<?php bloginfo( 'pingback_url' ); ?>"
/>
    <link rel="stylesheet" href="http://code.jquery.com/mobile/1.4.5/
jquery.mobile-1.4.5.min.css" />
    <script src="http://code.jquery.com/jquery-1.9.1.min.js"></script>
    <script src="http://code.jquery.com/mobile/1.4.5/jquery.
mobile-1.4.
5.min.js"></script>
        <?php wp_head(); ?>
  </head>
  <body <?php body_class(); ?>>
    <div data-role="page" data-theme="b" id="jqm-home">
      <div data-role="header">
        <h1><?php bloginfo('name'); ?></h1>
      </div>

      <div role="main", class="ui-content">
```

You probably recognize a lot of this code from previous projects, but there is a lot of new code here as well. So let's start breaking it down:

```
<!DOCTYPE html>
<html <?php language_attributes(); ?>>
  <head>
    <meta charset="<?php bloginfo( 'charset' ); ?>" />
    <title><?php wp_title(); ?></title>
    <link rel="profile" href="http://gmpg.org/xfn/11" />
    <link rel="pingback" href="<?php bloginfo( 'pingback_url' ); ?>"
/>
```

All WordPress themes must be written in correctly formed HTML5 and PHP. This means you must correctly open and close tags, divs, and so on. Here, we declare our page to be HTML5 and then use WordPress-specific PHP code to load in our WordPress site name, pingback URL, and more.

In the following code, we perform our standard loading of the jQuery Mobile framework via the jQuery CDN. We then call the WordPress header function and closing the `<head>` section of the code:

```
<link rel="stylesheet" href="http://code.jquery.com/mobile/1.4.5/
jquery.mobile-1.4.5.min.css" />
    <script src="http://code.jquery.com/jquery-1.9.1.min.js"></script>
    <script src="http://code.jquery.com/mobile/1.4.5/jquery.mobile-
1.4.5.min.js"></script>
<?php wp_head(); ?>
</head>
```

In the following snippet, we open the body of our HTML page and then our jQuery Mobile page and content data-roles:

```
<body <?php body_class(); ?>>
<div data-role="page" data-theme="b" id="jqm-home">
<div data-role="header">
  <h1><?php bloginfo('name'); ?></h1>
</div>

<div role="main", class="ui-content">
```

That is our `header.php`. Now on to the footer!

Creating the footer file

The `footer` file is going to be one of the easiest files to create in the whole project. Create `footer.php` and enter the following code in it:

```
</div><!-- data role content-->

<?php wp_footer(); ?>
  </div><!-- data role content-->
</body>
```

Told you it was going to be short! Nevertheless, let's break it down. We simply close some open `div` tags and then call the `wp_footer()` function.

Creating the function file

We've seen several WordPress functions throughout our files, such as `wp_header()` or `wp_footer`. In this section, we are going to look at how to hijack one of these functions for our use. Create a `functions.php` file and add the following code:

```php
<?php
function footer(){
   echo '<div data-role="footer" class="ui-bar"><a href="http://www.
packtpub.com" data-role="button" data-icon="info">Mastering jQuery
Mobile</a></div>';
} add_action('wp_footer', 'footer');
?>
```

What we do is create the standard website footer: the copyright section. We create a jQuery Mobile button that links to the Packt Publishing website. We then call the WordPress `add_action()` hook so we can make our `footer()` function the function that is executed whenever `wp_footer()` is called. Let's break that down further. Here is the usage of `add_action` from the WordPress codex:

```php
add_action( $hook, $function_to_add, $priority, $accepted_args );
```

For this example, we use the first two arguments: `$hook` and `$function_to_add`. Our `$hook` is `wp_footer`. There are several hooks you can use, which is beyond the scope of this book. Our second argument is `$function_to_add`, which in this case is the footer function we just created.

Creating the index file

Finally, we come to the final file in our theme, `index.php`. This is one of the main files of the theme and the second required file. We could technically cram every other file we made into this one file, but just because we can, it does not mean we should. Create the file and add this code to it:

```php
<?php get_header(); ?>

    <ul data-role="listview" data-inset="true" data-theme="c" data-
dividertheme="b">
        <?php if (have_posts()) : while (have_posts()) : the_post(); ?>
        <li><a href="<?php the_permalink() ?>"><?php the_title(); ?></
a></li>
        <?php endwhile; ?>
    </ul>
```

```php
<?php else : ?>
  <h2>Oh no there are no posts!</h2>
<?php endif; ?>


<?php get_footer(); ?>
```

Not necessarily a whole lot of code, but a lot that you probably haven't seen. So let's break it down a few lines at a time.

The following code snippet calls the `get_header()` function, which includes a lot of code needed for WordPress. You can override this if you want in your theme; if not, the default WordPress code will work:

```php
<?php get_header(); ?>
```

Now, we create a `listview` data-role that will contain all of our WordPress posts:

```html
<ul data-role="listview" data-inset="true" data-theme="c" data-
dividertheme="b">
```

In the following snippet, we check to see if we have any posts in the WordPress database, and if so, loop through them, adding them to our `listview` data-role. We then close the unordered list:

```php
<?php if (have_posts()) : while (have_posts()) : the_post(); ?>
    <li><a href="<?php the_permalink() ?>"><?php the_title(); ?></
a></li>
  <?php endwhile; ?>
</ul>
```

This is the closing section of loop, which executes the else block if there are no posts. Each fresh install of WordPress comes with a default post, so unless you delete it, you should not see this message.

And finally, there is `<?php get_footer(); ?>`, not to be confused with the `wp_footer()` action we coded in the `functions.php` file. This, just like the `get_header()` function, will execute some necessary code that you can override in your theme if you desire:

```php
<?php else : ?>
  <h2>Oh no there are no posts!</h2>
<?php endif; ?>
```

Seeing the theme in action

Now that we have created all the necessary files, we need to see the theme in action. Here is how you do that:

1. If you did not create the files in the directory we suggested at the beginning of this project, please do so now. The file was `wp-content/themes/jqmwptheme` in the WordPress `install` directory.

2. Once you copy your files over here, log in to the WordPress administration dashboard.

3. Once you log in and see the dashboard, mouse over the **Appearance** menu item on the left-hand side of the page and click on **Theme** from the fly-out menu.

4. Here, you should see every theme you have installed. If this is a fresh installation, you may only see a few. However, one of these themes you see should be **My Awesome jQuery Mobile Theme**.

5. Mouse over our theme and you will see an **Activate** option. Click on this option.

6. This will refresh the page and our theme will be the first one listed with the Active overlay on the bottom of the box.

7. Now click on the **Home** link at the very top of the screen and the non-admin view of the page will show up. You'll see something similar to the following screenshot on your screen:

Great work! It is very basic, but you now have a starting point to expand on for your own use.

If you wish to learn more about theming WordPress, we highly recommend checking out the WordPress Codex page at `https://codex.wordpress.org/Theme_Development`. Packt also has several books on WordPress and theming. We recommend starting with *WordPress Theme Development - Beginner's Guide*, by Rachel McCollin and Tessa Blakely Silver. You can find it at `https://www.packtpub.com/web-development/wordpress-theme-development-beginners-guide`.

Summary

In this chapter, we looked at how you can use jQuery Mobile with Node.js, RequireJS, and Backbone.js JavaScript libraries, and finally we created a mobile theme for the WordPress CMS system. The scope of the sections covered in this chapter is extremely vast and, hence, covering everything was not possible in this chapter. However, you should treat this as a crash course or a demonstration of how jQuery Mobile plays with other JavaScript libraries and frameworks. We encourage you to explore these libraries independently as well as along with jQuery Mobile.

In the next and the concluding chapter of this book, we will take a look at different best practices related to mobile development and also discuss how we can improve the efficiency of a jQuery Mobile website. This is going to be an interesting chapter and an extremely enriching one.

10
Mobile Best Practices and Efficiency

We have come a long way over the last nine chapters, covering various aspects of the jQuery Mobile framework. We started off by setting up the environment to work on a project using jQuery Mobile. We set up Aptana Studio and XAMPP to set up a small DB and host our PHP code. We looked at the various tools and techniques for responsive web design, such as RESS, Modernizer, and WURFL. We also looked at some testing tools, such as Screenfly. We discovered various jQuery Mobile widgets and used them to create a jQuery Mobile web application of our own. Further, we used Cordova to create a hybrid mobile application. All of this is cool, and we also have an application to show, but are we really done yet?

No. We are not done yet. In actual production environments, just having a pretty-looking, fully functional, bug-free application is just not good enough; it also has to be fast. A recent study revealed that an average user expects a web page on the mobile to load faster than on a desktop (or at least at the same speed); 64 percent of smartphone users expect pages to load in less than 4 seconds. If the page takes any longer to load, users might abandon the site completely or move on to do something else while the slow page still loads.

In this chapter, we will take a look at various performance optimization techniques that will help improve the page load time and improve the performance of your web application. We will also take a look at the various best practices that can be employed to optimize the images, CSS, HTML, and JavaScript so as to effectively improve the performance of any jQuery Mobile web application.

Best practices for design and layout

When you have decided to use jQuery Mobile as a platform for your mobile web application, there are a few things to keep in mind in terms of the overall design and layout. The following key pointers will help you better design your web application.

Icon size

Most of the newest smartphones feature the retina display, a screen that packs double the number of pixels into the same space as older devices. For designers, this immediately brings up the question, "How does this affect my current images? What do I need to make them look outstanding on the retina devices?" The answer is pretty simple, but let's understand the concept of retina display images first.

The basic concept of a retina image is that a larger image with double the number of pixels (for example, 60 x 60, twice the height and width) of the original image (30 x 30) will be set to fill half the space:

This can be done using the following code snippet:

```
<img src="star_icon_60x60" style="height: 30px; width:30px;">
```

You can similarly use the following media query or JavaScript solution:

- For media query:

```
@media only screen and (-webkit-min-device-ratio: 2){
    .retinaIcon{
        background: url(../images/star_icon_60x60.png);
        background-size: 50% 50%;
    }
}
```

- For JavaScript:

```
If(window.devicePixelRatio == 2){
  //Replace the image source by the new large image source.
  //Add background size if required
  //Add height width if required
}
```

Designing with Photoshop

When designing a mobile web application using tools such as Photoshop, always make sure that you are designing at double resolution. For example, the standard size for mobile devices is 320 x 480, but when designing, make sure you design on a 640 x 960 resolution. This is extremely beneficial to visualize how the design would look on an actual retina display. When presenting the design to the client, you can reduce the size of the canvas by 50 percent and put a device frame around the design; the clients will then be able to visualize the design as if viewing it on a retina display. Another advantage to this is that, when you design at double resolution, the icons that you create will also be created at double the resolution—the designers don't have to spend time creating icons for the retina display separately.

Fluid design

When designing for a mobile website, make sure that the layout lends itself to a fluid environment. Not all devices are 320 x 480 in size. 320 px is a good middle ground for many devices, but we have a wide spectrum of devices on the market with varying screen sizes and resolutions. You should consider what happens to elements placed side-by-side in smaller screen widths and what happens to the layout in the landscape mode. The design should flow efficiently from the portrait mode to the landscape mode as well.

Avoiding fixed footers or headers

There are a couple of reasons why the fixed footers and headers should be avoided on a mobile website.

When the device is in the landscape mode, the available screen area for the application reduces drastically. The available height is around 320 px. The minimum height for a header and footer is 40 px each. So out of the total available 320 px, we lose another 80 px to the fixed header and footer. This leaves only a visible area of 240 px for the user, which is too less and frustrating for the end user while using the application.

Consider another scenario where the user is using the application in landscape mode and the web page has some form fields on it. When the user taps on a particular form field, the device keyboard comes up, blocking all the visible area whatsoever.

There is another issue with the fixed footer. Consider a scenario where the web page has a number of form fields. The user taps on a form field. The onscreen keyboard is displayed. The fixed footer is displayed above the soft keyboard. This reduces the available screen area and visibility for the user. Furthermore, when the user is done filling the form field and the keyboard is gone, the fixed footer does not fall to the bottom of the screen. It just stays at the center of the screen. This is an unresolved issue for specific versions of Android and iOS reported with jQuery Mobile. Now there are a few JavaScript fixes for this issue—to fix the position of the footer programmatically. However, in such cases, you will notice that the footer fluctuates, which is very annoying for users and would never make it through QA. Many a time, you will be required to add some OS version–specific hacks to your code, which should be avoided as much as possible.

Avoiding tables

The use of tables should be avoided as far as possible on a mobile website or web application. This is because tables are not a very intuitive mode to display information on a mobile website. In the portrait mode, a maximum of two or three columns of data can be shown with a reasonable amount of data. Anything beyond three columns makes your design unacceptable for users.

jQuery Mobile does provide the table widget, where the user can select the columns he wishes to see. However, the end result is still the same. At the most, two or three columns are feasible on a mobile device.

Dialogs and popups

Dialogs and popups are a risky business on mobile websites. Consider the following before using dialogs or popups on a mobile website:

- The most important thing to consider before using dialogs is the amount of content that will be displayed within the dialog. If the content within the dialog is too long, then you are bound to face scrolling issues on certain versions of Android and iOS.

- You will notice on several occasions that on certain Android and iOS versions, the main page will also scroll with the dialog or popup, whereas on many other OS versions, only the content within the dialog or popup will scroll. This results in different behavior on different browsers and different OS versions, which will result in several QA issues. Even your project stakeholders will not agree to this inconsistent behavior.

- Let's assume another scenario, where the user is in the portrait mode, has opened a dialog, and now changes the orientation to landscape. The dialog or the popup takes some time to reposition to the center of the screen. This transition is not very smooth and will result in a poor user experience. Moreover, the popup might not reposition itself to the center of the screen and you might have to do so programmatically.

Forms

Be mindful of the input label placement on your forms. Some touch devices, such as the iPhone, zoom in to the form field when you tap on it. Displaying the label above the field makes a lot of sense in such a scenario. The label is visible to the user and provides better visibility as he/she scrolls down the page.

For form fields, you can include the label right inside the input field as a placeholder, which saves space on the mobile device screen. However, remember that not all mobile browsers support placeholder text. Moreover, there are chances that the users might lose context of the fields in a long form with multiple fields.

Best practices for images

Images are an integral part of any website and add a lot of flavor to it. E-commerce sites rely mainly on these images, and such sites just can't be slow when accessed on a mobile device. We will take a look at some ways to optimize images and effectively optimize website performance.

Don't use images at all

Yes, that is right, and yes, we are not crazy. You can make use of icon fonts and Scalable **Vector Graphic** (**SVG**) icons, which are scalable and can easily be customized. An icon font is basically a font file filled with icons and glyphs instead of the usual letters. To use the icon font, simply embed the icon web font, and you will be able to use any icon you need. SVGs are truly scalable vector graphics that you can use on the web.

Why do we need to use this, you ask. Well, the icon fonts and SVG icons solve a major problem for the mobile website. With the advent of multiple-dpi phones, designers have to build icons of different sizes to support the 2x and 3x retina display phones apart from the regular non-retina devices. Instead of maintaining multiple versions of the same icon, designers can make use of icon fonts or SVG icons, which will scale to all devices and not lose their clarity even on the retina devices.

Using icon font libraries such as Font Awesome or Pictos makes a lot of sense when your graphic is simple and in a single color. Use SVG icons if you need full color capability and don't need to support IE8. You have complete CSS control over both icon fonts and SVG icons. Using these will greatly reduce the size of your web page and, hence, improve the page load time.

> You can read more about Font Awesome (`fortawesome.github.io` — yes, the link is fortawesome and not fontawesome) and Pictos (`pictos.cc`) and integrate these in your existing jQuery Mobile projects, too.

Optimizing images

When using images is unavoidable, you can optimize the images that are being used by using one of the several tools that are available for image compressions. Image compression reduces the size of your images considerably, thus reducing the size of your web page. As a rule, first run the lossy optimizers and then the lossless. You should first make all the necessary changes to your images, such as resizing, cropping, or converting the images, and then use the image optimizers. Using image optimization tools is a great way to reduce the size of your image assets without compromising the quality of these images.

Image sprites

One of the main reasons for a slow loading web page is the multiple HTTP requests your browser must make to retrieve all of your web page; and the multiple images on this web page add to this problem. Multiple images on a web page means the browser has to make those many HTTP requests, which slows down the page.

To avoid this problem, make use of the image sprite. A sprite is an image that has all of your other images within it. So, if your sprite has five images, you effectively get rid of four HTTP requests, thereby speeding up your sites' loading time by the time for each of those four request's latency.

The same image set for retina and non-retina devices

Instead of maintaining two different sets of icons/images for retina and non-retina display devices, you can simply make use of images created for retina display devices and use them on non-retina display devices as well. This will reduce your code to conditionally check for retina or non-retina devices and display the right image accordingly.

The following code can be used to display a single image created for retina display devices on non-retina display devices as well:

```
.icon{
  background: url(../images/star_icon_2x.png);
  background-size: 50% 50%;
}
```

Now one can counter this by stating that we create a sprite image as discussed in the earlier section, and so, we do not make multiple HTTP requests for the different icon sets. However, if you just make use of the icons created for retina displays, you end up reducing the size of your sprite image as well, and this is extremely helpful to improve the performance of your web page.

Lazy loading

Lazy loading of images is used to defer the image requests to a point where it is absolutely necessary to load the images. Lazy loading is a very common feature on image-intensive sites, for example e-commerce and video streaming sites such as YouTube. Lazy loading helps reduce the page size, resulting in a faster loading page. It can be implemented fairly simply with the help of several JavaScript plugins that are freely available.

Best practices for CSS

When you are using a framework such as jQuery Mobile to build a website, you may not use all the components and widgets provided by this framework, which means that there is a lot of CSS that is never being utilized. To improve the performance of your website / web page, optimizing the CSS can prove greatly useful. Let's take a look at some of the CSS optimizations that we can implement to improve your page performance.

Customizing the jQuery Mobile download

You can customize your jQuery Mobile download by selecting the specific modules that you will need in your application. To download a custom build of jQuery Mobile, use the download builder available at `http://jQuerymobile.com/download-builder`.

Custom builds will reduce the library's file size. They also reduce the time taken by the library to spin up and start executing. All of this helps reduce the time taken to initialize the page. Let's take a look at a small example of a custom build, which includes the following:

- Full page and navigation support
- All custom events
- Slide transition
- Flip switch
- Checkboxes and radio buttons
- Text inputs
- Listviews
- Panels
- Popups
- Tables
- Column toggle tables
- Toolbars
- Fixed toolbars with workarounds
- Page loading widget

The custom build that contains these elements reduces the CSS size by a whopping 80 percent. This goes to show that if we create a custom build, we can reduce the library file size by a great extent.

Great! So why don't we just use a custom build all the time? Because there are a couple of disadvantages to this approach. For starters, jQuery Mobile's official documentation says that the jQuery Mobile download builder is still in alpha. As such, it should not be used on production websites. Secondly, using a custom build on production sites hampers the scalability of the website.

Let's assume that you have the preceding custom build, and now you want to make use of the datepicker widget. You will have to create a custom build again, which includes the datepicker widget, and only then will you be able to use it in your website. This increases the time taken to implement new features and your application's time to market.

Removing unused CSS

If you are working on a huge, complex website with a large team contributing to the CSS, then you are bound to end up with a lot of unused CSS selectors and rules. It is important to remove the unused CSS selectors and reduce the CSS file size to speed up the browser rendering your website. Before a browser can render your HTML page, it must download and parse any style sheets that are required to lay out the page. Once the style sheet is loaded, the browser's CSS engine has to evaluate every single rule contained in the file to see if the rule applies to the current page. There are several available free tools, such as mincss, which you can use to determine the unused CSS in your website, and with which you can remove or defer these unused CSS selectors and rules.

Combining multiple CSS files

We usually have multiple CSS files, which helps in organizing and maintaining the CSS during development. However, a separate HTTP request is required for each of these style sheets. A browser can download only a fixed number of files at a time, and this increases the load time of the web page. Combining multiple CSS files into one single style sheet helps improve the page load speed, as the number of HTTP requests is considerably reduced.

Minify and gzip

Minification is the process of removing all unnecessary characters from the source code without changing its functionality. The unnecessary characters usually include white space characters, new line characters, comments, and sometime block delimiters. All of these unnecessary characters help improve readability but are not necessary for code execution.

Gzip compression works by replacing similar text strings within a file and replacing these strings temporarily to make the overall file size smaller.

Minifying and compressing your CSS file greatly reduces the file size, which helps speed up the web page.

jQuery optimizations

We looked at various methods that we can implement to optimize images and CSS to improve the performance of web pages. In this section, we will take a look at some of the ways in which we can optimize our jQuery and JavaScript to enhance the performance of our web pages further. The first step to optimize JavaScript is to write clean and optimum JavaScript code. Assuming that we are writing the optimal JavaScript code for our web pages, let's take a look at some of the other things that can be done to optimize the code further.

Selector caching

You should *always* cache your jQuery selectors. It is extremely wasteful to constantly call $(selector) over and over with the same selector. If you are sure of using the same selector a number of times within your code, it makes complete sense to cache this selector. This improves the page speed considerably. Let's take a look at a simple code snippet to understand how you can cache your selectors:

```
var cache = $(this).parent().parent().parent().children().siblings()
[1];
$(cache).slideUp('slow');
```

Using this query multiple times would be very costly and, hence, caching it to a variable makes more sense.

Script files at the end or in the head?

What is the ideal location to include script files in your document? This is always open for debate. It is often argued that for speed purposes, you should include the script files right at the end of the document just before the closing body tag. This means that the script files will not interfere with the loading of the rest of the HTML document. This will result into fast page loading, which seems to be a great advantage. However, it can result in poor user interaction.

Let's take a simple example:

```
$(document).ready(function(){
  $("li:nth-child(odd)").addClass("odd-row");
})
```

Here, the `odd-row` class applies some background color to that particular list item. The application of this effect will often be visible to the user, which makes for a bad user experience. You are better off including your script files in the head section of your page if your users don't get disconcerting visual effects.

Combining the JavaScript files

We usually have multiple JavaScript files, which helps in organizing and maintaining the JavaScript code in development environments. However, a separate HTTP request is required for each of these files. A browser can download only a fixed number of files at a time, which increases the load time of the web page. Thus, combining multiple JavaScript files into one file helps improve the page load speed as the number of HTTP requests is considerably reduced.

Using the latest jQuery version

The core jQuery team is always working on improving the library through better code readability, introducing new functionality, and optimizing existing algorithms. For this reason, always use the latest available version of jQuery. You can link to the latest jQuery using the following line of code.

```
<script src="http://code.jquery.com/jquery-latest.js"></script>
```

However, there is a downside to this. It may create problems in the future, when the latest version of jQuery is served to your web application even before you have had the chance to test your site using the newest version. So instead of linking to the latest version of jQuery, you can link to the latest stable version of jQuery that you need:

```
<script src="https://code.jquery.com/jquery-1.11.2.min.js"></script>
```

jQuery Mobile custom build

Let's consider the same example of the custom build that we evaluated in the CSS enhancement techniques section. If the custom build is to include the same widgets as in the earlier example, it reduces the JavaScript size by a whopping 50 percent. This goes to show that if we create a custom build, we can reduce the library file size to a great extent. But this approach has its own pitfalls, as discussed in the earlier section.

Don't always use jQuery

A number of times, the plain old vanilla JavaScript is easier and faster than jQuery. Let's consider an example.

We wish to change a few CSS properties of an HTML element. Here's how we go about it using jQuery:

```
$(selector).css({
    backgroundColor: 'red';
    color: '#FFF';
});
```

```
Using plain Javascript, the same code becomes much simple and faster!
    var text = document.getElementById(selector);
    text.style.backgroundColor = 'red';
    text.style.color = '#FFF';
```

Let's consider another example of a very common function used for looping. The use of native JavaScript functions is nearly always faster than the ones in jQuery. For this very reason, instead of using the `$.each` method, we should make use of JavaScript's native `for` loop. Also, there is another advantage of using the native `for` loop. We can cache the length of an array/collection, which helps speed up things.

Consider the following `for` loop, where during each iteration, the length of the array will be recalculated:

```
for (i=0; i<array.length; i++){
    //Your logic goes here
}
```

Instead, we can use the following `for` loop, where the length of the array is calculated and saved, and then used during each iteration:

```
for (i=0, len = array.length; i<len; i++){
    //Your logic goes here
}
```

Minify and gzip

Minification is the process of removing all unnecessary characters from the source code without changing its functionality. The unnecessary characters usually include white space characters, new line characters, comments, and sometime block delimiters. All of these unnecessary characters helps improve readability but are not necessary for code execution.

Gzip compression works by replacing similar text strings with a file and replacing these strings temporarily to make the overall file size smaller.

Minifying and compressing your JavaScript files greatly reduces the file size, which helps speed up the web page.

jQuery Mobile optimizations

When using a framework such as jQuery Mobile, there is a lot of scope for optimization, as frameworks provide a number of things out of the box that may not be utilized but are still part of the web page. This increases the page load time and thereby affects the performance. We will take a look at some optimization techniques that will help us improve the performance of our jQuery Mobile web applications.

A multipage template

You can consider making use of the multipage template in jQuery Mobile in certain cases where there is small number of pages in the web application. Using a multipage template can improve the performance of a jQuery Mobile web application, but it also has its downsides. Before taking a look at the pros and cons of using a multipage template, let's try to understand the page structure in jQuery Mobile.

The page is the primary unit of interaction in a jQuery Mobile web application. An HTML document may start with a single page, and the Ajax navigation system loads additional pages on demand inside the DOM as the users navigate through the site. In a single-page template, only one HTML page will be loaded into the DOM at any given time, and previous pages will be removed from it unless specified otherwise.

A multipage template on the other hand is basically a collection of all the pages loaded as part of a single HTML file into the DOM. All pages are loaded in the DOM and are available at all times, which results in faster page transitions; but such web applications are bloated.

Using a multipage template is an option only if your application has very few pages. This is a very good option if you are creating a simple static website for mobile devices. Using a multipage template has the following advantages:

- All pages load with a single HTTP request
- Pages load faster
- Page transitions are fast and clean

Though using a multipage template improves page-loading speed, it is not always a recommended approach due to the following limitations:

- The initial page download is extremely slow. This is because the page size is heavy since all the pages are available as part of one single HTML.
- The multipage template leads to a very large DOM depending on the size of the web application (number of pages).
- It uses a lot of system memory, which is limited on mobile devices.
- The jQuery Mobile library cannot load pages via Ajax in a multipage template.

Prefetching pages

When using the single-page template, the jQuery Mobile library allows us to prefetch other pages into the DOM so that they are readily available to the user when he visits these pages. To prefetch a page, we need to add a `data-prefetch` attribute to the link that is pointing to the page that needs to be fetched. jQuery Mobile then loads the target page in the background after the primary page has been loaded and the `pagecreate` event has been triggered:

```
<a href="page2.html" data-prefetch="true">Navigate to page #2</a>
```

This can also be done programmatically using the `pagecontainer` widget's `load()` method:

```
$(":mobile-pagecontainer").pagecontainer("load", pageUrl,
{showLoadMsg: false});
```

Server-side processing for single-page templates

We saw that the multipage template has several disadvantages, and that is a reason why using the single-page template is more popular for web applications. This does not mean that the single-page template is a flawless solution. It also has its own disadvantages. In a single-page template, everything included in the page HTML is sent every time the page is requested. However, when a page is loaded via Ajax, all that is required is `div` with `data-role="page"`. This means a lot of unused information is sent with every request. This unused information leads to extra processing time when the page is being loaded. To optimize the single-page template, we can add a little bit of server-side processing. We can check whether the page is being loaded using a regular HTTP request or an Ajax request. If it is an Ajax request, then we are going to send `div` with `data-role="page"`.

Let's take a look at how we will be doing this:

```php
<?php
  if(!isset($_server['HTTP_X_REQUESTED_WITH']) || strtolower($_
server['HTTP_X_REQESTED_WITH']) !== 'xmlhttprequest') {
?>
  <!DOCTYPE html>
  <html>
    <head>
      <title></title>
      <meta name="viewport" content="width=device-width, initial-
scale=1">
      <link rel="stylesheet" href="http://code.jquery.com/
mobile/1.4.5/jquery.mobile-1.4.5.min.css" />
      <script src="http://code.jquery.com/jquery-1.11.1.min.js"></
script>
      <script src="http://code.jquery.com/mobile/1.4.5/jquery.mobile-
1.4.5.min.js"></script>
    </head>
    <body>
    <?php } ?>
    <div data-role="page">
      <div data-role="header">
        <h1>Page Title</h1>
      </div>
      <div role="main" class="ui-content">
        <p>Page content goes here.</p>
      </div>
      <div data-role="footer">
        <h4>Page Footer</h4>
      </div>
    </div>
    <?php if(!isset($_server['HTTP_X_REQUESTED_WITH']) ||
strtolower($_server['HTTP_X_REQESTED_WITH']) !== 'xmlhttprequest') {
    ?>
    </body>
  </html>
<?php } ?>
```

When we execute this code, we end up reducing the HTTP request to send only `div` with `data-role="page"`:

```
<div data-role="page">
  <div data-role="header">
    <h1>Page Title</h1>
```

```
    </div>
    <div role="main" class="ui-content">
      <p>Page content goes here.</p>
    </div>
    <div data-role="footer">
      <h4>Page Footer</h4>
    </div>
  </div>
```

This reduces the request size considerably. With jQuery Mobile 1.4 onwards, we can use widgets outside the page as well. Taking advantage of this, we can exclude the common parts of HTML, such as the header and the footer, and further reduce the HTTP request to send only the following:

```
<div data-role="page">
  <div role="main" class="ui-content">
    <p>Page content goes here.</p>
  </div>
</div>
```

This effective updated template reduces the size of the HTTP request. It also reduces the markup that needs to be parsed. Since we also excluded common widgets such as the header and footer, we also made sure that these widgets are not re-initialized with every request. This saves a lot of time, and the web page loads much faster.

Pre-enhanced markup

This is a very powerful way to improve the load time of a jQuery mobile web page. Before we drive into the details of this approach, let's try and understand how jQuery Mobile works.

Let's take an example of the collapsible-set widget. We use the following code to include a collapsible-set widget on our web page:

```
<div data-role="collapsible">
  <h4>Heading</h4>
  <p>Some text content comes here</p>
</div>
```

When jQuery Mobile encounters data-role="collapsible" on a div element, it will look for a header (h1-h6) or a legend element immediately following this div tag, and it will style this header to look like a clickable button and add a + icon to the left to indicate that it is expandable. The entire HTML that follows will be wrapped within a container div and will be shown or hidden when the heading is clicked.

This clearly means that the jQuery Mobile framework spends time to instantiate the widget on the page as well as to enhance it. Considering that there are a number of widgets on the web page, a great amount of time is spent on initializing and enhancing each of these widgets on the web page. This makes the web page slower; thus, it will load extremely slowly on a mobile device that does not have good hardware and has a slow data connection.

We can reduce this time taken by jQuery Mobile to enhance the widgets so as to improve the load time of our web page. When the framework enhances a particular `div` based on its `data-role`, it adds a certain set of classes to the markup and a few new elements in the DOM. What if we can do this?

If we can do this, then the framework will not have to spend time enhancing a widget, which will improve the page load time considerably. The question is can we do it?

The answer is yes. The jQuery Mobile framework has a provision to render pre-enhanced markup without re-initializing the same widget. By providing the markup required for the collapsible-set widget, we can instruct the framework to skip all DOM manipulations during instantiation and to assume that the required DOM is already present. We can do this by setting the `data-enhanced="true"` attribute.

Setting the `data-enhanced="true"` data attribute is not enough. You have to make sure that you also set all the classes that the framework would normally set along with all the data attributes whose values differ from the default values.

In the following example, pre-rendered markup for a collapsible has been provided. The `data-collapsed-icon="arrow-r"` attribute is explicitly specified, since the use of the `ui-icon-arrow-r` class on the anchor element indicates that the value of the `collapsedIcon` widget option is to be `arrow-r`:

```
<div data-role="collapsible" data-enhanced="true" class="ui-
collapsible ui-collapsible-inset ui-corner-all ui-collapsible-
collapsed" data-collapsed-icon="arrow-r">
  <h2 class="ui-collapsible-heading ui-collapsible-heading-collapsed">
    <a href="#" class="ui-collapsible-heading-toggle ui-btn ui-btn-
icon-left ui-icon-arrow-r">
      Title
      <span class="ui-collapsible-heading-status"> click to expand
contents</span>
    </a>
  </h2>
  <div class="ui-collapsible-content ui-collapsible-content-collapsed"
aria-hidden="true">
    <p>Content within the collapsible comes here</p>
  </div>
</div>
```

Using data defaults

This is an extremely useful option to improve the page load time when using the default options for a particular jQuery Mobile widget. Let's take a look at the advantage of setting data defaults to true. However, before we proceed to take a look at this feature, let's take a look at what options are available for different widgets in jQuery Mobile.

There are several different options available for all the widgets available in jQuery Mobile. These options deal with the presentation of a particular widget. Options such as `corners`, `iconpos`, `inset`, `mini`, and so on, deal with various UI aspects of a particular widget. For example, corners will set rounded corners on the widget, and `iconpos` will inform the widget of the position of the icon within that widget, whether the widget should be the regular size or a mini variant, and so on.

All these options have certain default values. For example, the `corners` option is always set to `true` by default as of the current version. The `mini` option is set to `false` by default, which means that the widget is of a regular size by default.

Now that we understand what options a widget offers, we will be able appreciate the value that the defaults option brings to the table.

Setting the defaults option to true indicates that the other option values of the widget have default values and causes jQuery Mobile's widget auto-enhancement code to omit the step where it retrieves option values from the data-attributes. This improves the page load time considerably if you have a number of widgets on your web page, all of which make use of the default option values. This option is exposed as a data attribute: `data-defaults="true"`. By default, this option is set to `false`.

You can also initialize a widget, for example the collapsible widget, with the defaults option specified as follows:

```
$(selector).collapsible({
  defaults: true
});
```

Number of widgets on a page

When we include various widgets on a web page, the jQuery Mobile library applies the enhancements to the DOM elements by initializing each widget. A significant amount of time is spent on initializing each widget and applying the enhancements to these HTML elements. So, if we have fewer widgets included as part of a single page, we can reduce the time necessary to initialize each of these widgets and thereby improve the page load speed.

Limiting the size of widgets

When building a mobile web application, we tend to include a very long list of items, since we can leverage the infinite scroll on a mobile device. However, including a long list of listview items and a large number of rows in tables is a huge kill for the performance of the jQuery Mobile application. We should avoid having a really long list of listview items, and instead, make use of pagination or lazy loading techniques. This helps reduce the number of list items available in the DOM at a given moment, which improves the page load time, thereby improving the performance of your jQuery Mobile web application.

Performance tools for optimization

In this chapter, we have so far covered various methods to optimize our code, that will help the page load time and thereby improve the performance of our mobile web application. In this section, we will take a look at some of the tools and utilities that we can use to leverage performance enhancements.

In this section, we will not look at code optimizations, but we will look at how we can leverage the different browser developer tools and some interesting browser plugins that will be of help. We will also consider how we can debug our mobile web code using remote debugging techniques for both Android and iOS devices. Finally, we will take a look at the Google PageSpeed Insights tool. Let us now take the final step towards optimizing our mobile web application.

The Google Chrome developer tools

When working on the development of any web page, the Google Chrome developer tools come in very handy. The developer tools not only allow you to inspect and edit the DOM elements or edit the CSS on-the-fly, they also bring several other benefits with them. One of the most important features that Google Chrome developer tools provides for mobile web development is Device Mode and Mobile Emulation.

When we write code for our mobile web application, the work is only half done. We have to still test it, and here we are faced with a problem. Testing web pages on different devices becomes a complex and time-consuming task. Moreover, often of times, we do not have various devices available with us to test our web pages, and it's here that the Chrome developer tools come to the rescue.

The device mode in Google Chrome brings the insights of mobile testing to your desktop browser by emulating the mobile experience, as shown in the following screenshot:

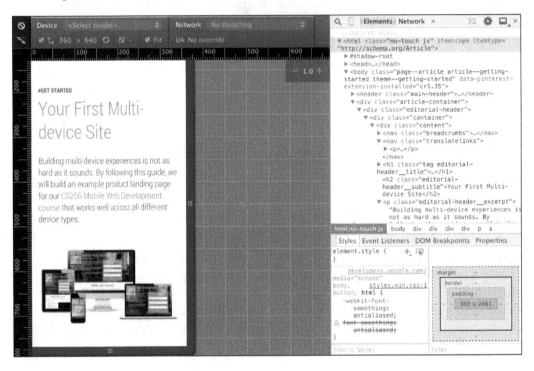

The device mode can be enabled by clicking on the small mobile device icon at the top-left corner of the Chrome developer tools console. The device mode allows you to do the following:

- It allows you to test the responsiveness of your mobile web pages by emulating different screen sizes and resolutions of a mobile device, and select retina device emulation.
- The network emulator allows you to evaluate the performance of your web page.
- The styles section enables visualization and inspection of CSS media queries.
- In this mode, you can accurately simulate device input for touch events and geolocation, and most importantly, you can also simulate device orientation.

Another important feature of the Chrome developer tools that has proved to be very important is the Network Performance Emulator:

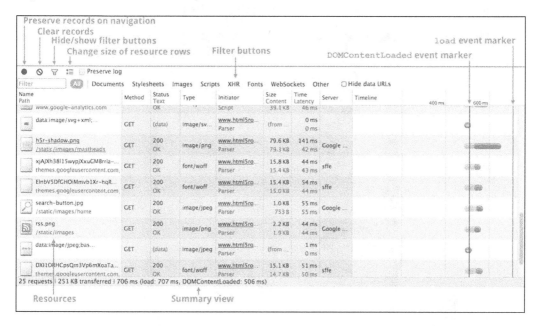

The **Network** panel in the Google Chrome developer tools records information about every network operation in your web application. This includes detailed timing data, HTTP request and response headers, cookies, WebSocket data, and much more. The **Network** panel helps us understand how much time a resource is taking to load, or where a particular network request was initiated.

The Google Chrome developer tools provide many more useful features that you would be interested in, which would definitely prove helpful to you for web page development. We strongly encourage you to explore Chrome's developer tools further at their official documentation website at `https://developer.chrome.com/home`.

Firefox tools and the Firebug plugin

For people who prefer using Firefox as their development tool, they should most definitely make use of the Firebug plugin for Firefox. You can download this super awesome tool from `http://getfirebug.com/`. Like the Google Chrome developer tools, Firebug helps you inspect and modify your HTML DOM elements and CSS styles in real time. It also provides an advanced JavaScript debugger tool and a network panel that will help you accurately analyze network usage and performance.

You can also test the responsiveness of your mobile web pages by making use of the inbuilt emulator. Just press *Ctrl + Shift + M* to resize the viewport of your browser to emulate various mobile device sizes and even the device orientation:

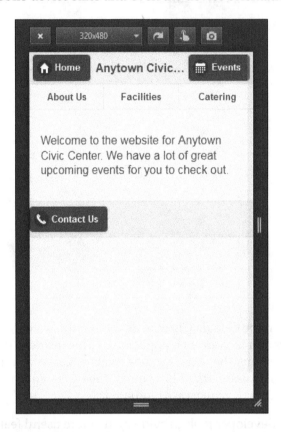

Remote debugging on Android

How many times have you come across issues on your mobile web project that are specific to some particular Android devices or some particular Android version, and on the Android stock browser or the installed webkit browser, or the Chrome browser? Most of you must have been in a situation like this where things are working fine everywhere except on some Android version. That is the time when you must have thought, "If only we could debug on the mobile browser just as we do on the desktop browser using our developer tools."

The Chrome developer tools will help you debug remotely over USB as long as an Android mobile device is connected to your development machine. You can then inspect and view the HTML, JavaScript, and CSS styles until you get a bug-free web page:

We encourage you to use the remote debugging technique effectively to help you improve the quality of your mobile web pages. You will find the details at `https://developer.chrome.com/devtools/docs/remote-debugging`.

Remote debugging on iOS

You can leverage remote debugging not only on your Android devices but also on your iOS devices. The process is very similar with some slight obvious differences pertaining to iOS devices. Apple added the ability to debug applications remotely in iOS 6. To debug a mobile web page on an iOS device, you need to have the device connected to your Mac development machine via USB and the remote debugging option enabled on your iOS mobile device:

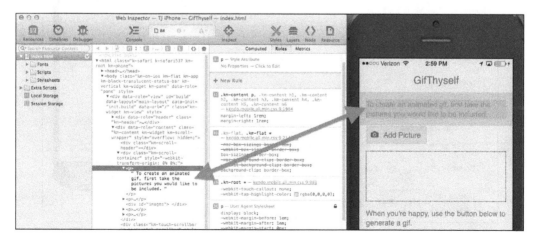

We encourage you to use the official documentation on iOS-Safari remote debugging and make use of this technique to enhance your mobile web pages and make them bug-free.

The Google PageSpeed tools

The PageSpeed tools are a set of another awesome and very useful tools for web developers. These tools are designed to help developers optimize the performance of a website. The PageSpeed optimization tools, when applied to a mobile website, evaluate the page speed of the site and also suggest best practices that can be applied to a website to further enhance the page speed and optimize it.

Fast and optimized pages lead to higher visitor engagement and retention and better conversion rates. Google PageSpeed tools are available as a Google Chrome browser plugin. You can even analyze your site online. In the following screenshot, you can see how the PageSpeed tools analyze Google's website performance for both the mobile and desktop websites. The PageSpeed tools then suggest some changes that can be made to further enhance the page speed of the website:

Summary

In this chapter, we went beyond just the jQuery Mobile framework and focused on the different web page optimization and performance enhancement techniques. We evaluated why page optimization is important and how it affects the user experience on the mobile web. We evaluated several different best practices for designing and layout for the mobile web, so that even before we start with website development, we can have a good responsive design in place. Once a good design is in place, we can develop the website in a better fashion and make use of the various HTML, CSS, jQuery, JavaScript, and jQuery Mobile best practices to enhance and improve the quality of our code. This helps in improving the performance of the mobile website. Developing good quality code is just not sufficient. Testing it is equally important, and we explored a few tools that can help us with this to enhance our website performance further.

With this, we come to the end of this book. We hope that you have thoroughly enjoyed and benefitted from our small effort. We have tried our best to bring value to this book and help you develop a liking for, and to master, an awesome jQuery-based framework: jQuery Mobile.

Index

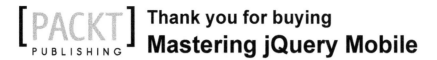
Thank you for buying
Mastering jQuery Mobile

About Packt Publishing

Packt, pronounced 'packed', published its first book, *Mastering phpMyAdmin for Effective MySQL Management*, in April 2004, and subsequently continued to specialize in publishing highly focused books on specific technologies and solutions.

Our books and publications share the experiences of your fellow IT professionals in adapting and customizing today's systems, applications, and frameworks. Our solution-based books give you the knowledge and power to customize the software and technologies you're using to get the job done. Packt books are more specific and less general than the IT books you have seen in the past. Our unique business model allows us to bring you more focused information, giving you more of what you need to know, and less of what you don't.

Packt is a modern yet unique publishing company that focuses on producing quality, cutting-edge books for communities of developers, administrators, and newbies alike. For more information, please visit our website at www.packtpub.com.

Writing for Packt

We welcome all inquiries from people who are interested in authoring. Book proposals should be sent to author@packtpub.com. If your book idea is still at an early stage and you would like to discuss it first before writing a formal book proposal, then please contact us; one of our commissioning editors will get in touch with you.

We're not just looking for published authors; if you have strong technical skills but no writing experience, our experienced editors can help you develop a writing career, or simply get some additional reward for your expertise.

Creating Mobile Apps with jQuery Mobile
Second Edition

ISBN: 978-1-78355-511-6   Paperback: 288 pages

Create fully responsive and versatile real-world apps for smartphones with jQuery Mobile 1.4.5

1. Learn how to integrate advanced features such as Geolocation, HTML 5 Video, and the Web Audio API into your web application.

2. Enhance your efficiency by automating repetitive tasks with the Grunt task runner.

3. Effortlessly blend and integrate jQuery Mobile with existing WordPress, Drupal, and HarpJS projects.

Mastering jQuery UI

ISBN: 978-1-78328-665-2   Paperback: 312 pages

Become an expert in creating real-world Rich Internet Applications using the varied components of jQuery UI

1. Create useful mashups by plugging together different components along with APIs.

2. Design your own widgets like captchas, a color picker, news reader, puzzles, and many others.

3. Take your jQuery UI skills to next level by exploring the ins and outs and nuances of jQuery UI components.

Please check **www.PacktPub.com** for information on our titles

Mastering jQuery

ISBN: 978-1-78398-546-3   Paperback: 400 pages

Elevate your development skills by leveraging every available ounce of jQuery

1. Create and decouple custom event types to efficiently use them and suit your users' needs.

2. Incorporate custom, optimized versions of the jQuery library into your pages to maximize the efficiency of your website.

3. Get the most out of jQuery by gaining exposure to real-world examples with tricks and tips to enhance your skills.

Functional Programming in JavaScript

ISBN: 978-1-78439-822-4   Paperback: 172 pages

Unlock the powers of functional programming hidden within JavaScript to build smarter, cleaner, and more reliable web apps

1. Discover what functional programming is, why it's effective, and how it's used in JavaScript.

2. Understand and optimize JavaScript's hidden potential as a true functional language.

3. Explore the best coding practices for real-world applications.

Please check **www.PacktPub.com** for information on our titles